The radiation-induced decomposition
of inorganic molecular ions

The radiation-induced decomposition of inorganic molecular ions

EVERETT R. JOHNSON

University of Maryland

GORDON AND BREACH SCIENCE PUBLISHERS

New York   London   Paris

Copyright © *1970 by:*

Gordon and Breach, Science Publishers, Inc.
150 Fifth Avenue
New York, N.Y. 10011

Editorial office for the United Kingdom:

Gordon and Breach, Science Publishers Ltd.
12 Bloomsbury Way
London W.C.1

Editorial office for France:

Gordon & Breach
7–9 rue Emile
Paris 14e

Preface

There is considerable interest in the radiation induced decomposition of inorganic molecular ions primarily because of the many interesting complex solid state reactions that occur in these systems. For the most part, the reported research only reveals the stoichiometry of these complex reactions and leaves much to be done in terms of our understanding of the mechanism. Much of the reported results contain controversial data or theories i.e. only in a few cases do we find agreement among the different laboratories. In some instances the only information that exists is that reported by a single laboratory. More importantly, however, there have been a number of isolated, and significant observations reported which have not been verified or fully developed. Consequently, it appeared more appropriate to render due recognition to all contributions rather than attempt to edit the whole. Admittedly some editing has been done, however, this has been kept at a minimum. Where possible, an attempt is made to resolve controversial data or at least indicate lack of internal consistency. What has been attempted here is to produce a "state of the art" manuscript that will permit an easy entre into the field.

Chapter 1 provides a brief survey of those topics of solid state chemistry which are necessary for an understanding of what follows. The intent is to introduce the reader to the vocabulary rather than give precise discussion. In Chapter 2 we have discussed those aspects of inorganic decompositions which are believed to be general phenomena; i.e. back reaction, thermal decomposition, energy transfer etc. and to a large extent its content consists of the edited ideas of the author. The subsequent chapters deal with the research reported on specific classes of compounds.

This manuscript evolved as a result of the author's research, which was wholly sponsored by the Division of Research, U.S. Atomic Energy Commission. The author gratefully acknowledges this support and hopes that this manuscript will act as a catalyst for increasing research in this fascinating field.

EVERETT R. JOHNSON

v

Contents

1

Effects of high energy radiation on solids

1.1   Introduction

In this monograph we are primarily concerned with the chemical consequences (in the broadest sense) of the interaction of ionizing radiation with those crystalline inorganic solids which have been characterized as containing molecular ions. Such solids as the nitrates, bromates, chlorates, azides, etc., fall in this category while those such as the alkali and other halides do not.

Much of the research reported assumes some knowledge of the physics and chemistry of the solid state. Accordingly, in this chapter we present a necessarily brief review of those aspects of the solid state which are very pertinent to an understanding of much of the research discussed.

1.2   General

The interaction of ionizing radiation in these systems is complex and is usually described on the basis of: (1) primary interactions which involve electronic transitions of the molecules producing ions, excited species, radicals, etc., and in certain circumstances atom displacements; and (2) secondary interactions which usually involve the reactions of the ions, excited species, radicals, etc., so produced to give the final products. In most cases only the final products (stable) are actually observed and the nature of the intermediate or primary species, which are the precursors of the final products, is for the most part speculative.

The number of primary (excluding atom displacement processes) events occurring along the path of an individual ionizing particle is usually described in terms of Linear Energy Transfer or LET. This is simply the mean energy loss of the ionizing particle per unit of path-length and is usually reported in terms of eV/Angstrom. For example the LET for a 3.4 MeV alpha particle is about 34 eV/Å whereas that for cobalt − 60 gamma rays is about 0.06 eV/Å.

The effect of LET is studied to provide insight into the mechanism of the reaction. High LET will cause an increase in the local population density of all intermediate species and hence an increased probability of certain reactions which may or may not effect the final product distribution. High LET in solids, however, also produces high local temperatures which causes a local expansion (or even local melting) along the track of the ionizing particle. The effect of high LET in solids has often been compared to the effect of temperature on the radiation induced decomposition of these salts, but as will become apparent in later sections this comparison is not always valid.

The chemical changes produced during irradiation are related to the energy absorbed by the term "absorbed dose". A convenient notation that follows naturally from the concept of the quantum yield characterizing the absorbed dose is the G value or 100 eV yield. This quantity is the number of molecules decomposed, or formed, or reacted per 100 eV of energy absorbed. More generally it is the number of any specie (ion, radical, molecule, etc.) formed, reacted, or decomposed per 100 eV absorbed. The absorbed dose may be measured by a variety of techniques; however, the most common method (and the most reliable) is the oxidation of acid ferrous sulfate solutions (Fricke Dosimetry). A complete description of this and other methods of determining absorbed dose is given in references 1 and 2.

Radiation effects in solids may be viewed as producing defects and these defects in turn may be classified as primarily atom displacement or an altered electronic configuration. These two categories of damage are by no means independent of each other, since an altered electronic configuration will certainly affect the position of neighboring atoms and, of course, vice versa. However, in some materials, metals for example, radiation damage is primarily atom displacement since changes in electronic configuration are rapidly adjusted.

1.3   Nature of defects in solids

There is a large variety of defects in solids that are normally present and which have been characterized. The basic structures of these defects may be categorized as point defects and dislocations. Several subcategories give to each a particular identity or name. Point defects include lattice vacancies and interstitial atoms. Dislocations are a volume or linear defect in which the lattice is disturbed a few lattice distances away radial to the dislocation. Frenkel and Schottky defects are point defects and refer to interstitial and vacancy pairs respectively.

1.3.1   Schottky defects

A perfect ionic lattice (we assume an AB type lattice in these discussions) has an equal number of anions and cations. A Schottky defect may be envisaged as an ion leaving a normal lattice site and taking up a position on the surface, leaving behind a positive or negative ion vacancy. The number of

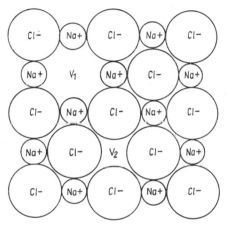

FIGURE 1-1   Schematic of Schottky defects. $V_1 = $ anion vacancy, $V_2 = $ cation vacancy

cation and anion vacancies are always equal (in the absence of impurities). If an excess number of cations migrate to the surface, a positive charge is produced thus preventing further migration of positive ions, and at the same time creating excess negative charge within the crystal. This latter effect is conducive to migration of the negative charge, so that in the absence of external forces the number of oppositely charged vacancies inside a crystal tends to become equal (Figure 1-1). Schottky defects are therefore vacancy pairs.

Schottky defects are always present in a solid. In any normal pure solid, formed under equilibrium conditions, there exists an equilibrium number of these vacancies which depends on the temperature at which the crystal has been formed.

1*

At equilibrium for a given temperature T the number of Schottky defects is given by:

$$\frac{N - n}{n} = \exp\left[Gp/kT\right] \tag{1-1}$$

where Gp = Gibbs energy required to create a vacancy pair
$\quad\quad k$ = Boltzman constant
$\quad\quad N$ = no. of atoms in crystal
$\quad\quad n$ = no. of Schottky defects

since N is usually $\gg n$ equation (1) may be simplified to give

$$n \cong N \exp\left(-Gp/kT\right). \tag{1-2}$$

1.3.2 Frenkel defects

These are characterized by the movement of an ion or atom to an interstitial position, as shown schematically in Figure 1-2 forming an interstitial vacancy pair. Like Schottky defects, the Frenkel defects are always present,

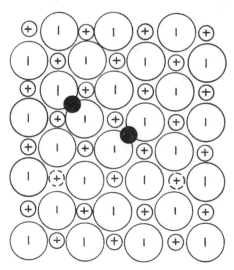

FIGURE 1-2   Schematic of Frenkel defects

their number being a function of the temperature at which the crystal is formed. The formula expressing the number of Frenkel defects is:

$$n \cong (NN_i)^{1/2} \exp\left(-G_f/2kT\right) \tag{1-3}$$

where N = number of atoms in the crystal

$\quad N_i$ = number of interstitial positions

$\quad G_f$ = Gibbs energy required to form an interstitial pair.

It is seen that even in the absence of radiation there always exists a finite number of vacancies and interstitials and it is therefore virtually impossible to obtain a crystal which does not contain these defects.

It must be construed that the defect concentration produced at the high temperatures is present in the crystal at room temperature. Diffusional and other processes rapidly reduce the high temperature concentration during the cooling process. Defects are "annealed" by a wide variety of processes. Gibbon and Kuczynski,[3] for example, have shown that defects induced in alkali halides when the salts are pressed into pellets are rapidly annealed at interfaces at room temperature. Formulas (1-2) and (1-3) obtained on the bases of equilibrium thermodynamics can only be expected to give very approximate results for normal crystals.

1.3.3 Color centers

Another type of defect in solids is that manifested by color centers. Most organic and inorganic compounds become colored when exposed to ionizing radiation. In some cases the coloration produced is caused by the formation of a new stable compound or relatively stable radical that is absorbing; or the color may be due to the formation of an ion such as the naphthalenide ion that occurs when diluted solutions of naphthalene in 2-methyltetrahydro-furan are irradiated in the glassy state. The term "color center", however, is usually reserved for a loosely bound electron or hole trapped at a vacancy or other point defect and which occurs primarily in ionic, insulating, and semiconducting lattices. The term is used in the literature for radiation induced coloring in all substances, but will be used here in a more restrictive sense: the introduction of localized states in the forbidden gap which act as electron or hole traps.

As indicated above any crystal grown under equilibrium conditions contains defects and dislocations. The presence of such defects alters the charge distribution, and a change in the electronic levels in the vicinity of the defects should be expected. Prior to the general use of ionizing radiation, color centers were produced in alkali halide lattices by the introduction of excess alkali metal or halogen. If alkali metal is deposited on the surface of an alkali halide crystal, a coloration is produced in the crystal. This occurs presumably by an anion migrating through the lattice, reacting with the metal atom, and the liberated electron being trapped in the anion vacancy. Such crystals show an absorption band in the visible region (KCl, dark blue; LiF, pink; NaCl, brown). This model was confirmed by the fact that the absorption produced by adding excess metal was independent of the metal used, i.e., addition of sodium metal to KCl produced the same absorption band as if potassium had been used. The model has also been confirmed by ESR techniques; the splitting observed is 0.1995 compared to 2.0023 for the free electron.

The name given to this type of color center is F center (from the German "farbe"). When the F center is irradiated by white light, the trapped electron can be excited to higher-lying quantum states. The wavelength of light that can excite these trapped electrons is not singular, and it is found that a range of wavelengths can be used; hence an absorption band is found which is called an F band. When such a crystal is irradiated with light in the F band region and an electric field is applied to the crystal, photoconductivity is observed.

When excess halogen is added to an alkali halide crystal (NaCl, for example), new color bands are produced. These new bands, called V bands, are formed presumably by halogen atoms, occupying normal lattice sites, which produce a positive hole in the lattice and these holes are trapped. Such color centers are complex; electron spin resonance has shown that some of these centers are not spherically symmetrical. The V_k center is a hole trapped at a halide ion in the perfect crystal. Chemically speaking, this may be thought of as a Cl_2^- ion. These centers are stable only at low temperatures. The H center is a hole trapped at an interstitial halide ion and may be thought of chemically as an interstitial halogen atom. This center also is stable only at low temperatures.

Some other color centers are the U center, which is an impurity center and consists of a hydride ion substituted for a normal halide ion, and the α center, which is an ionized F center and is thought to arise from the transfer of an electron from a neighboring halide ion to an alkali ion. More than a dozen color centers have been identified in KCl alone. Figure 1-3 is a schematic representation of some color centers.

Color centers in substances other than the alkali halides have been reported but the characterization of these centers is far from complete. Where applicable, these will be treated in the ensuing sections under the appropriate heading (nitrates, chlorates, etc.) The characterization of color centers in the more complex crystals is often facilitated by the application of ESR, and thermal and optical bleaching techniques.

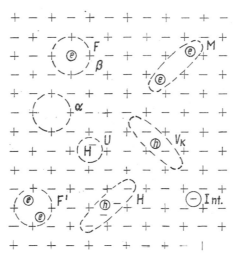

FIGURE 1-3   Schematic of some color centres

Table 1-1 gives the approximate wavelengths of the *F*-band centers in some alkali halide crystals.

The number of color centers may be determined approximately by using an equation originally dervied by Smakula but simplified by Przibram.[4]

TABLE 1-1   Approximate wavelengths of *F* band

Alkali halide	Position of max in Angstroms	Half-width eV at 20°C	Oscillator strength
NaF	3400	0.62	—
NaCl	4650	0.47	0.81
NaBr	5400	—	—
NaI	5880	—	—
KF	4550	0.41	—
KCl	5630	0.36	0.90
KBr	6300	0.36	0.80
KI	6850	0.35	—
RbCl	6240	0.31	—
RbBr	7200	0.28	—
RbI	7750	—	—

The formula can be applied only if the band has a Gaussian or Lorentzian shape.

$$fd = 1.31 \times 10^{17} \frac{n}{(n^2 + 2)^2} \alpha_{max}H \qquad (1\text{-}4)$$

where f = oscillator strength

n = index of refraction at λ_{max}

d = concentration of absorbing centers per cm^3

α_{max} = absorption coefficient in cm^{-1} at λ_{max} where λ_{max} is the wavelength at peak absorption

H = half-width of the band in eV.

In order to obtain the color center concentration it is first necessary to plot the optical density versus wavelength and determine the shape of the band, and λ_{max}. The absorption coefficient α is equal to the product of the exinction coefficient and the concentration, $\alpha = \varepsilon C$, hence the optical density at λ_{max} divided by the path length gives α, the absorption coefficient at λ_{max}. The oscillator strength must also be known; however, these generally vary between 0.6 and 1. For a single emission electron there is a sum rule which gives a maximum value of 1; hence, using an approximate value for f, a rough value of the color center density can be obtained.

1.3.4   Other defects

Impurities are also defects in a lattice. In a chemically reactive system, impurities build up with exposure to radiation. For example, irradiation of KNO_3 produces NO_2^- and O_2, both of which are impurities and may act as trapping sites; etc.; i.e., new or additional defect centers. Impurities may also be formed by nuclear mutation. All these defects may be capable of forming color centers or be directly related or involved in the decomposition process.

1.3.5   Chemical detection of colour centers

The detection of color centers by chemical means has been attempted in several instances. Chemically speaking, trapped electrons and holes should behave in the alkali halides as free alkali atoms and free halogen atoms. When the irradiated salt is dissolved in aqueous solution, the "electrons" should react to produce hydrogen (act as a reducing agent) and the holes act as an oxidizing agent. Rabe, Rabe and Allen[5] used acidic solutions of ethanol and determined the hydrogen gas evolved. Burns and Williams[6] dissolved the irradiated crystals in H_2O and determined the hydrogen yield and the hypochlorite formed. In both instances, the yield of hydrogen was greater than the F center concentration as calculated from equation (1-4). Hacskaylo and Otterson determined the "F" center yield by pH determination and "V" center yield assuming the reaction to be equivalent to the reaction of hypochlorous acid with O-tolidine.[7]

Heal[8] used liquid ammonia and saturated aqueous solutions of mercuric chloride to detect free electrons (or sodium atoms) in irradiated azides. The reactions are

$$e + NH_3 \longrightarrow [e(NH_3)]^-$$
$$e \text{ (or Na)} + H_2O \longrightarrow OH^- + H$$

the hydrogen atoms reacting with mercuric chloride to produce insoluble mercurous chloride.

1.4   Production of defects

Energy may be absorbed in a solid by (1) electron displacement (ionization, excitation), (2) atom displacement, (3) production of an impurity atom or molecule, and (4) electron trapping. The energy adsorption process occurs in about 10^{-17} sec. This may be followed by molecular dissociation, luminescence, and so forth, which occurs in about 10^{-13}–10^{-9} sec. Phosphorescence, diffusional processes, chemical reactions, color center decay, hole pair interactions and the like complete the time scale of events to restore equilibrium, which may be achieved in from 10^{-8} sec up to times as large as years.

Radiation effects in solids are broadly characterized as those which are primarily crystallographic structural defects, and those which are chemical in nature. The predominant type of damage will depend to a large measure on the material (metals, ionic solids, semiconductors, and such) and on the source of radiation. Heavy particle radiation (fast neutrons, fission fragments, fast protons, and so on) may cause atom displacement and chemical change while photons, and electrons produce mostly chemical changes, if they can occur.

1.4.1   Interaction of neutrons with solids

Fast neutrons interact primarily by direct collision with nuclei (knock-on collision). The result is a displaced atom which is ionized and behaves as a heavy charged particle, causing some ionization and other displacements. Slow neutrons react almost entirely by capture processes to produce impurity (defect) atoms which may possess sufficient recoil energy to cause ionization and excitation and induce displacement.

The source of fast neutrons is usually a reactor since machine sources in general do not produce a sufficiently high flux for their practical use in most damage studies. Reactor neutrons are always accompanied by gamma rays and the sample therefore is invariably exposed in a reactor to the presence of a gamma field. For a chemically reacting system the effect of decomposition by fast neutrons (via displaced atoms) is usually small compared to the decomposition induced by the gamma field accompanying fission neutrons. The average number of displacements produced by fast neutrons may be determined approximately by equation (1-5):[9]

$$N \approx \frac{E}{4T_m} \left[0.108 - 0.561 \log \frac{E_\tau}{E} \right] \left[1 + \frac{E_\tau}{E} \right]^{-1} \qquad (1\text{-}5)$$

where E = the energy of the neutron
E_τ = the energy required to displace an atom
T_m = the maximum energy of the primary displaced atom.

The value of E_τ is estimated for most solids to be about 10 eV,[10] which is large compared to the energy to produce a Frenkel defect. This is because the recoil takes place before the neighboring atoms have a chance to relax. T_m the maximum energy of the primary knock-on, is given by

$$T_m = \frac{4EMm}{(M + m)^2} \qquad (1\text{-}6)$$

where m = mass of the neutron
M = mass of the struck atom.

1.4.2   Heavy charged particles (alpha, proton, etc.)

For energetic charged particle interactions the collision is coulombic. These particles lose energy by ionization and excitation and also by atom displacement. The maximum energy transferred by atom displacement is given by equation (1-6), which is also valid for a hard sphere collision. The mean energy transferred to the primary atom is:

$$\bar{E} = E_\tau \ln \frac{(T_m)}{E_\tau} \quad \text{for} \quad T_m \gg E \qquad (1\text{-}7)$$

where the symbols have the same meaning as in Equation (1-5).

Equation (1-7) gives the mean energy of the primary recoils. The average number of secondaries dislodged from the lattice by the primary atom is:

$$N = \left[0.885 + 0.561 \ln \frac{(X_m + 1)}{4} \right] \frac{X_m + 1}{X_m} \qquad (1\text{-}8)$$

where $X_m + 1 = \dfrac{4Mm}{(M + m)} \dfrac{E}{E_\tau}$

M = mass of struck atom

m = mass of particle.

For example, bombarding copper with 500-keV deutrons, $\bar{E} = 195$ eV and $\bar{N} \approx 4.5$.

1.4.3   Electrons and gamma rays

Irradiation with energetic electrons (~ 1–2 MeV) or energetic gamma rays (> 1 MeV) can produce displacements. An outstanding example is the disordering produced in copper gold alloy by ^{60}Co gamma rays. The number of displacements produced however are so small, compared to chemical damage (in a chemically reacting system), that displacement damage relative to other effects may be ignored. The primary atoms displaced by energetic electrons will have very little energy and hence secondary production is virtually nil. Irradiation with electrons or photons will cause primarily excitation and ionization and in chemically reacting systems, decomposition.

1.4.4   Fission fragments and recoils

If a fissionable substance is incorporated into a lattice and the material then exposed to a neutron source, the fission fragments produced in the lattice are essentially massive-charged particles of high energy which can produce ionization and atom displacements; i.e., extensive damage may be produced depending upon concentration, etc.

Mutation resulting from nuclear reactions are a very common source of recoils. Hot-atom chemists employ this means to study the chemical effects of nuclear transformations. The literature on this subject is substantial.*

---

* See for example G. Harbottle *Radioisotopes in the field of Recoil Chemistry*, IAE, Vienna (1962).

The elements commonly used in many of these studies, particularly in organic systems, are lithium (which becomes a triton) and nitrogen (which becomes carbon-14).

1.4.5   Other mechanisms for producing defects in solids

As previously indicated, fast neutrons, heavy charged particles, fission fragments or recoils resulting from slow or fast neutron capture produce defects in crystals. These defects include atom displacements, chemical decomposition, and color centers. Irradiation with photons or electrons, on the other hand, would result primarily in an altered electronic configuration (chemical decomposition, color centers, radicals, and so forth) and very little or no atom displacement. There are several additional mechanisms which have been postulated for the production of atom displacement in crystals.

1.4.5.1   Thermal spikes   There are basically two types of thermal spikes to be considered. One is that caused by a knock-on atom (primary recoil) of energy E, with the deposition of this amount of energy in a small, restricted volume.

This sudden deposition of energy may result in very high local temperatures for a short duration of time. For example, using an energy deposition of 10^5 eV in a volume containing about 10^4 atoms and a thermal diffusivity of 10^{-3} cm^2 per second, this number of atoms could be heated to temperatures of about 1000°K in a time duration of about 10^{-11} second. The role of this type of thermal spike can only be significant in heavy particle or neutron bombardment.[9]

1.4.5.2   Electron spike   The other type of thermal spike is that which is sometimes called an electron spike. This is the spike associated with the passage of a charged particle (electron) where the condition exists that little or no energy is lost directly to the lattice, the charged particle energy being dissipated entirely to the electrons and the electronic excitation energy subsequently being transferred (degraded) as heat. The rate at which energy is converted from electronic excitation energy to lattice energy will decide whether sufficiently high lattice temperatures can be achieved to cause change. The magnitude of the energy transferred will depend directly on the degree of coupling between the electronic states and the lattice. For metals where the coupling is very weak, Seitz and Kohler[9] show that lattice temperatures will be of the order of 500°K for 10 eV electron excitation per atom along the track of the ionizing particle. This temperature is only a hundredth of the electron temperature, and is not considered significant in terms of lattice changes; however, when the coupling is much larger, Seitz and Kohler believe that electron spikes may be significant.

In terms of chemical change produced, conversion of electronic excitation energy to lattice vibration energy may be quite significant. However, there is as yet no definite theoretical or experimental evidence to support the view that this event plays an important role in the chemical change observed in radiation induced decomposition of inorganic or organic systems. It can only be said that this remains a possibility.

1.4.5.3   Varley mechanism   Another method of producing defects is that postulated by Varley[11] which applies to ionic solids. Multiple ionization of an anion to produce a net positive charge will result in the anion finding itself in a highly unstable position by virtue of the crystal potential (it will be surrounded by positive charges). Multiple ionization could occur as a result of an Auger cascade following readjustment of the electron cloud initiated by inner-shell ionization. In the case of neon, for example, multiple ionization will occur in 16 percent of those atoms losing a K electron. Because of this unstable condition the ionized anion could be forced out of its equilibrium position to an interstitial one. The Varley mechanism, however, would be most likely to occur only with simple anions such as the halides; with polyatomic anions it would be expected that decomposition of the anion would occur.

1.4.5.4   Inner-shell ionization   Durup and Platzman[12] have developed methods for calculating absolute yields for inner-shell ionization. The results indicate that the G value for inner-shell ionization falls rapidly when the initial electron energy is less than one hundred times the threshold energy for the pertinent event. At initial electron energy of about 1 MeV, G_k values for K shell ionization for Li, F, K, and Cl are 0.155, 0.00681, 0.00027 and 0.000373 respectively. In fluorine, G_L values for ejection of an electron in the L shell is calculated as 0.3; for potassium and chlorine the L and M subshell yields are significantly larger than K shell yields (the order of 20 to 100 G_K).

1.4.5.5   Other mechanisms   Although the Varley mechanism is still thought to be valid, a more recent mechanism[13] is the formation of excited molecular ions (X_2^{2-} in alkali halides) whose subsequent breakup gives a vacancy and interstitial ion. This latter mechanism is rapidly gaining acceptance.

1.5   Properties sensitive to radiation damage

For substances of most interest to chemists, the properties of principal concern are thermal, electromagnetic, physical, and electrical properties. Mechanical properties are, of course, definitely affected in irradiated solids, but these are of minor importance in inorganic and organic crystals. A significant mechanical effect in complex inorganic crystals is a general destruction of the crystal forces producing changes in density and, in those cases where significant chemical change has occurred, pulverization. In some simple salts such as LiF, hardening may occur and the yield stress may increase (in some cases by as much as a factor of 2).

1.5.1   Thermal properties

The presence of a large amount of point defects will cause a decrease in the thermal conductivity. This is caused by the scattering of phonons. Phonons are vibrational quanta associated with the equilibrium oscillations of an atom (or a complex of atoms) to produce lattice waves. Heat is transferred through a crystal by phonons and, if there exists a large number of defects, the phonons will be scattered, producing a decrease in the thermal conductivity.

Stored energy in the form of defects in crystal lattices is an important thermal property. Frenkel defects, for example, require the expenditure of about 2 to 5 eV; hence, these defects represent stored energy in a crystal. A particular example of this is the stored energy due to defects in carbon moderators in nuclear reactors. If this energy should be released suddenly, serious damage may result.* The stored energy in crystals (which may represent a variety of defects) of interest to the chemist can be easily determined by heats of solution measurement. Modern techniques for this type of study will detect at least 0.01 cal per gram of stored energy.

1.5.2   Electromagnetic properties

The electric potential is altered in the vicinity of point defects and these defects then may act as electron scattering centers which produce an increase in electrical resistivity.

The effect on the electrical potential by the defect often results in the introduction of localized electronic states producing optical absorption bands (F, F', V centers, and so on) which are manifested by a coloring of the

---

* This was the basic cause of the severe reactor accident which resulted in the closing down of the reactor at Windscale, Lancashire, England.

crystal, luminescence and photoconductivity. As indicated previously, the determination of the species that gives rise to a color center is often very difficult to resolve for complex molecules, and may require a combination of EPR (or NMR), optical spectroscopy, radical scavenging (isotopic labeling), thermal annealing, optical bleaching, or other techniques for identification.[13a]

1.5.3   Electrical properties

There are several electrical property measurements which can be made and which are of value; however, of these there are only two which are tractable; these are conductivity (or resistivity) and photoconductivity. These measurements give a very good- indication of the number of free-charge carriers (indicative of the number of electrons which escape the parent positive ion), the number and energy distribution of electron trapping sites, depth of traps, and the ability of the free-charge carriers to transfer energy to the surrounding molecules.

Photoconductivity is the conductivity observed when a sample is illuminated with light (in a strict definition this could be extended to irradiation with X- or gamma-rays). By studying the conductivity as a function of temperature, wavelength, and combinations of these, valuable information can be obtained concerning the centers responsible for the supply of conduction electrons. This technique is often used in conjunction with ESR.

1.5.4   Other properties

In almost all cases of irradiated inorganic materials that undergo chemical decomposition (including alkali halides), a change in density is observed. Such a change in density is often accompanied by changes in X-ray pattern and infrared absorption curve. X-ray studies of irradiated inorganic crystals will, in general, show line broadening, and if the crystal is nearly perfect, general distortion of peak reflections. Infrared studies on irradiated crystals may show broadening of the principal bands and the production of many side bands which are probably due to general distortion of the lattice planes. Raman spectra have also been of value especially in the identification in some instances of a new phase.

There may also be changes in properties such as solubility, melting point, and surface characteristics. Any property or physical measurement that is dependent on diffusion in the solid may be affected because diffusion in crystals is sensitive to the defect concentration.

References

1   J. W. T. Spinks and R. J. Woods, *An Introduction to Radiation Chemistry*, John Wiley and Sons, New York 1964

2   E. J. Henley and E. R. Johnson, *The Physics and Chemistry of High Energy Reactions*, Chapt. 4, Plenum Press (1969)

3   C. F. Gibbon and G. C. Kuczynski, *J. Chem. Phys.* **46**, 814 (1967)

4   K. Przibram, *Irradiation Colours and Luminescence*, Pergamon Press, London (1956)

5   J. G. Rabe, B. Rabe, and A. O. Allen, *J. Phys. Chem.* **70**, 1098 (1966)

6   W. G. Burns and T. F. Williams, *Nature* **175**, 1043 (1955)

7   M. Hacskaylo and D. Otterson, *J. Chem. Phys.* **21**, 552 (1953), IBID 1434 (1953)

8   H. G. Heal, *Trans. Far. Soc.* **53**, 210 (1957)

9   F. Seitz and J. S. Koehler, in *Solid State Physics*, F. Seitz and D. Turnbull eds. vol. 2 Academic Press, N. Y. (1954)

10 Y. Querre, in *Action Chimique et Biologiques des Radiations*, M. Hassinsky ed. Masson, Paris (1964)

11 J. H. O. Varley, *Nature* **174**, 886 (1954)

12 J. Durup and R. Platzman, *Disc. Far. Soc.* **31**, 156 (1961)

13 D. Pooley, *Color Center Conference*, U. of Ill. (1965)

13a J. Cunningham, *Int. Jour. of the Phys. and Chem. of Solids* **23**, 843 (1962)

2

Factors affecting the decomposition of molecular ions

The information on the decomposition of molecular ions has now developed to the point where some general ideas concerning the radiation induced decomposition of inorganic solids have gained acceptance. By this is simply meant that sufficient information has been obtained so that certain parameters in these decompositions which are not unique to a particular class of compounds may be identified. The parameters which will be discussed here are: role of lattice orientation, energy transfer, back reaction and thermal decomposition.

2.1 Crystal environment

The role of lattice parameters in the decomposition of inorganic salts was first indicated by Hennig, Lees and Matheson.[14] They observed a very large difference in the decomposition yield (G values) of KNO_3 and $NaNO_3$ (about 1.5 and 0.2 respectively) and did not feel that the small difference in bond energies could account for this and so postulated that the difference in the decomposition yield of homologous salts could best be explained by the difference in "free space". The free space is defined as the difference between the actual volume of the crystal per ion pair and the combined volume of the ions as calculated from standard ionic radii. Although some studies appeared to lend support to this concept, there were serious discrepancies. For example, in the nitrates, $CsNO_3$ has a higher decomposition yield than KNO_3 yet it has a much smaller free space; barium nitrate has a free space comparable to that of sodium nitrate but has the largest yield of any of the nitrates studied. There are other more subtler discrepancies. For example, oxygen diffuses out of the $CsNO_3$ lattice much more readily than it does from the KNO_3 lattice yet back reaction ($NO_2^- + \frac{1}{2}O_2 \rightsquigarrow NO_3^-$) is more pronounced in this salt than in the more loosely bound lattice of KNO_3. There is no correlation between free space in the radiolysis of the alkali and alkaline perchlorates[15] if all salts are included in the plot. A correlation between free space and initial *G*-values for the alkali bromates is observed,[16] but this correlation does not hold if the alkaline earth bromates are included in the plot.[17] No correlation between free space and the decomposition of the chlorates has been found.[18] Furthermore, these studies showed that the ratios of the yield of the products were identical in all the salts studied indicating that the relative probability for a particular decomposition mode was independent of the cation.

Free space, per se, does not appear to be a determining factor in the radiolysis of a homolgous series of salts; it is just a particular manifestation of the important parameter which we will call the total crystal environment.[19]

We may also add here that no correlation of decomposition yields with lattice energies, densities, melting points, or free energies of formation has been found with nitrates or perchlorates. Doigan and Davis[20] have postulated that the field strength exerted by the cation on the anion is a significant parameter in the photochemical decomposition of the nitrates; however, this does not hold for the radiation induced decomposition. Baberkin[60] has expressed a similar idea for the radiolysis of the nitrates, but only limited agreement is found.

2.1.1   Role of lattice orientation

If appears from most of the evidence accumulated to date that the mode (product distribution) and extent of decomposition of both organic and inorganic substances are often dependent on the spacial arrangement of the ions in the lattice. This would be the orientation of the ion or molecule in the lattice, and would not necessarily have any relation to "free space".

Irradiation of crystalline choline chloride at room temperature results in decomposition with *G* values as high as 55,000. The decomposition may be expressed as:

$$(CH_3)NCH_2CH_2OH^+Cl^- \rightsquigarrow (CH_3)_2NH^+Cl^- + CH_3CHO.$$

At about 73–78°C choline chloride undergoes a phase transition from orthorhombic to face-centered cubic. Irradiation at this temperature gives a very significant decrease in product yield.[21] Irradiation of alkyl halides in the glassy state produces significant differences in product yields than when the identical material is irradiated in the polycrystalline form.[22] The polymerization of n-alkyl-n-vinylsulfonamides induced by ionizing radiation occurs with a much higher yield in the solid than in the liquid. Most spectacularly, irradiation at $-75°C$ induces polymerization in the crystalline state, but not in the glassy state.[23] Spatial orientation has been used to explain the observed dimer products in the radiolysis of solid hexane.[31]

Barium bromate as purchased is the mono-hydrate, this salt decomposes during irradiation to yield BrO_2^-, BrO^- and Br^- (and possibly some BrO_2). Heating this salt up to about 180°C produces marked changes in the X-ray spectrum indicating new phase or phases and a substantial change in the radiation induced decomposition yield.[24]

Potassium nitrate undergoes two phase transition at the triple point 125.8°C.[25,26,27] These phases are labelled II, I and III. Above 125° phase I exists, on slow cooling instead of going back to phase II, the normal room temperature phase, a new phase, phase III is formed, which is ferromagnetic. Phase II is reformed only after the crystal has been cooled to about 110°C. Below this temperature there is only one phase. Irradiation of KNO_3 at 122°C and 190°C show a marked change in G values.[28] If the salt had been heated slightly above 122°C and then cooled to 122°, it is quite likely that some fraction of the salt had undergone this transition to phase III. These results have been substantiated by Hochanadel[68] who has shown that the yield in KNO_3 radiolysis is doubled when the decomposition is done at 150°. No change in decomposition yield is observed when KNO_3 is irradiated at 80°C.

$CsNO_3$ undergoes a transition at about 154°C, return to equilibrium is very slow, hence heating this salt to the transition temperature and then cooling to some lower temperature will leave some of the high temperature phase. Hochanadel has found a substantial increase in yield when this salt is irradiated at 150°C. Rubidium nitrate is another salt which does not show any appreciable change in G values with temperatures until the salt is irradiated above its transition temperature. $RbNO_3$ undergoes a phase transition at 164°C; irradiation at 122°C shows a slight increase in yield, however, irradiation at 190°C produces a substantial increase in yield. Lithium nitrate undergoes no phase transitions and this salt shows no change in G value when irradiated at 150°C. Sodium nitrate has a very substantial temperature coefficient for the radiation induced decomposition.

2*

Although this salt undergoes no phase transition it does exhibit some unusual properties when heated. For one it undergoes a marked volume change— so much so that it can be observed optically. In addition there are substantial changes in the specific heat with increased temperature.

Annealing experiments in irradiated nitrates also demonstrates the effect of lattice orientation. For example heating KNO_3 below the phase transition temperature shows very little annealing (back reaction of the products, nitrite ion and oxygen to form KNO_3) whereas heating above this temperature shows a very dramatic increase.[29] Numerous other examples may be found where a change in phase has resulted in very marked changes in behavior when the material is exposed to ionizing radiation.[30]

It is apparent therefore, that the orientation of the molecules in the lattice is an important and significant parameter in the radiation chemistry of these systems. It is our opinion that the complete crystal environment is the principal factor governing variation in yields in a homologous series of salts. Factors such as free space, cation size, lattice energies, etc. are variables of the total crystal environment and are probably *all* involved in some complex manner in the decomposition of these salts.

2.2   Energy transfer

In pure crystalline solids there is strong coupling of the electronic states, and an excited state should not be considered as confined to a single molecule.* This is in contrast to a molecule in the gas phase, solution or rigid glasses. In these systems the energy levels are essentially those of an isolated molecule. In rigid glasses such as boric acid, and in frozen solutions as for example those of aromatic hydrocarbons, fluorescence when exhibited shows no change in lifetime from that observed in the gas phase or liquid solution.

Energy transfer occurs by two basic processes; electronic and thermal. The transfer of electronic excitation may take a variety of forms depending on the system. Some mechanisms are unique to solids such as exciton transfer and the transfer of electronic excitation energy to lattice vibrational energy. Thermal effects are also different in solids because of the lack of mobility of the components of the lattice.

---

* This is not true of deep lying levels such as f states of rare earths.

2.2.1 Excitons

The transfer of energy by excitons is limited entirely to semiconducting or insulating crystals. This would include molecular crystals, ionic crystals, ceramics, most polymers, semiconductors, and so forth. Excitons are the excited electronic states in these crystals. The exciton is in principle an electron hole pair, and may be considered as a conduction-band electron and valence-band hole bound together but with separation, the pair traveling through the crystal. The movement of the exciton through a crystal may be envisioned as occuring by a process of recombination and subsequent reabsorption of the liberated energy; the excited molecule finds itself surrounded by others that are capable of accepting and reemitting the excitation. The exciton may wander throughout the crystal until some event occurs such as interaction with phonons (scattering) or with impurity atoms, dislocations, or point defects, which results in a loss of the exciton energy or it may recombine without emission of radiation and the excitation energy transferred to the lattice in the form of heat. In those cases where the exciton has interacted with impurity atoms or defects it can no longer wander and remains trapped and localized; the energy of the exciton being dispersed by fluorescence decay to the ground state or even chemical decomposition of the species where trapping has occurred.

The principal contribution of exciton-type transfer of energy is the ability to transport energy to another site far removed from the initial point of absorption. The importance of exciton transfer in crystals may be studied by observing the effect of the addition of well characterized impurity atoms or of increasing the defect concentration. For example, in a system containing 1 part in 10,000 of tetracene in anthracene, the exciton becomes trapped on the tetracene because the base of electronic state of tetracene is lower than that of anthracene. The energy of the exciton is emitted as fluorescence from the tetracene molecule. In irradiated pure inorganic crystals no luminescene (except in certain phosphors) is observed, the excitation energy (unless there is interaction with defects) being disapated as heat. When an exciton recombines the heat generated may be quite large; in KCl for example this is about 6 eV.

The energy of excitons varies with the nature of the crystal. In a semi-conductor the ground state of the crystal corresponds to an integral number of valence bands completely full and separated by an energy gap from completely empty conduction bands. In the excited states electrons are excited across the gap. Exciton energies can be determined from the absorption spectra of the solid. In ionic lattices band-to-band transitions are estimated to occur approximately 1 eV beyond the first absorption peak.

In crystals a sufficiently energetic photon or ionizing radiation (electrons, etc.) can create a free electron-hole pair and these particles can travel independently through the conduction and valences bands. The electron hole pair subsequently recombine at defects (hole traps, recombination centers, etc.). Energy transfer by electron migration is distinguished from exciton transfer as follows: in exciton transfer we think of a rapid transfer of energy by the exciton wave packet to some particular trapping site (if such exist in the crystal) and the significant event occuring; electron migration on the other hand is thought of as a particular molecule in the solid being ionized and the "liberated" electron migrating to some "trapping" site where the significant event occurs.*

2.2.2   Evidence for energy transfer in inorganic solids

There is abundant evidence in the literature for energy transfer in inorganic solids.[32] When KNO_3, embedded in a KBr matrix is irradiated, the KNO_3 undergoes decomposition which is far in excess of that which occurs in the pure salt[33] (when the energy absorbed by the KNO_3 is calculated on the basis its electron fraction in the total system). In this case energy absorbed by the KBr is transferred to the KNO_3 with subsequent decomposition of the KNO_3. When azoethane and other substances adsorbed on silica are irradiated it is observed that the decomposition of the adsorbate is very large compared to that of the pure material (again the energy absorbed by the absorbate is calculated on the basis of its electron fraction in the total system). The evidence for energy transfer in these systems is substantial. The mechanism of the energy transfer, however, has not been delineated. Evidence exists for electron migration as it does for exciton transfer.

Recently Khare and Johnson[34] have investigated energy transfer in systems composed of nitrates and some halates imbedded in various matrices. Figure 2-1 gives a $G(KNO_2)$ as a function of absorbed dose in this system for various concentration of KNO_3 in a KBr matrix. As can be readily seen $G(KNO_2)$ falls off rapidly with absorbed dose. It is also evident that $G(KNO_2)$ is a strong function of KNO_3 concentration. The critical para-

---

* See for example *Excitons* D. L. Dexter and R. S. Knox *Interscience Tracts on Physics and Astronomy no.* **25,** Interscience New York (1965).

meter in these systems is "absorbed dose" i.e. beyond a certain absorbed dose the matrix material is not able to transfer further energy and the G value falls off sharply. The remarkable coincidence is that the absorbed dose at which this effect occurs appears to be independent of the matrix material.

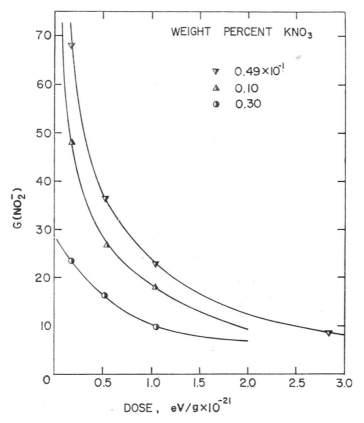

FIGURE 2-1   Decomposition of KNO_3 embedded in KBr matrix

The value found in these studies is about $0.5–1 \times 10^{21}$ eV/g. This is about the same dose observed by Allen *et al.*, Heintz *et al.*[32] and others in heterogeneous systems when SiO_2, MgO, and some alkali halides are used as inert supporters; i.e. at this absorbed dose G (decomposition) saturates for the adsorbate. This effect is shown in Figures 2-2 and 2-3. As can be seen it becomes constant after an absorbed dose of approximately 1×10^{21} eV/g.

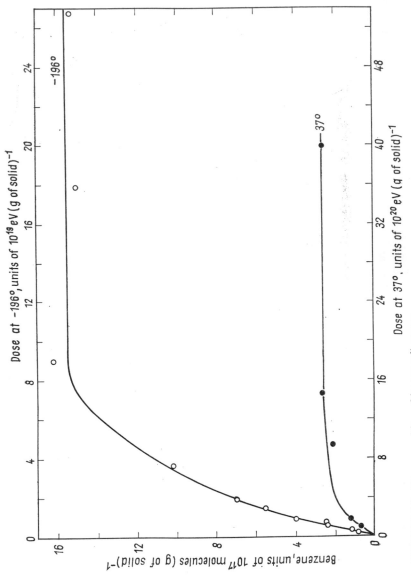

FIGURE 2-2   Decomposition of isopropylbenzene

In all these cases defect centers of some type which have been produced by irradiation have been postulated as preventing further energy transfer. Rabe, Rabe and Allen[5] determined the F center concentration in alkali halides and indicated a correlation between the decrease in decomposition of the adsorbate (for a silica-azoethane system) with the increase in F center concentration (Figure 2-4). In the system studied by Khare and Johnson the

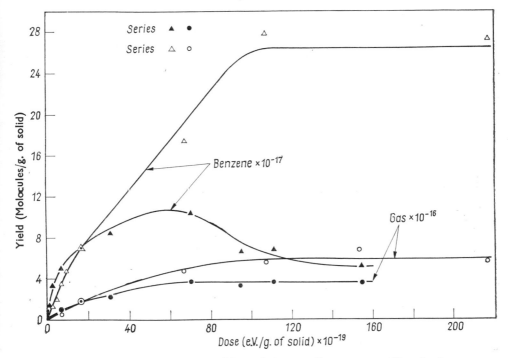

FIGURE 2-3 Decomposition of isopropylbenzene on silica–alumina.
Series 1 = ▲ ●, Series 2 = △ ○

nitrate ion was dissolved in the lattice and in effect could be represented as an impurity in the alkali halide matrix. G values for nitrate decomposition in these systems was as high as 250 which is more than two orders of magnitude greater than G values for the pure component. These studies were not sufficiently definitive regarding the mechanism of the energy transfer however the major evidence supports the concept of "exciton transfer" rather than electron transfer. Sagert and Robinson[35] studied the decomposition of N_2O adsorbed on silica gel and Zirconia

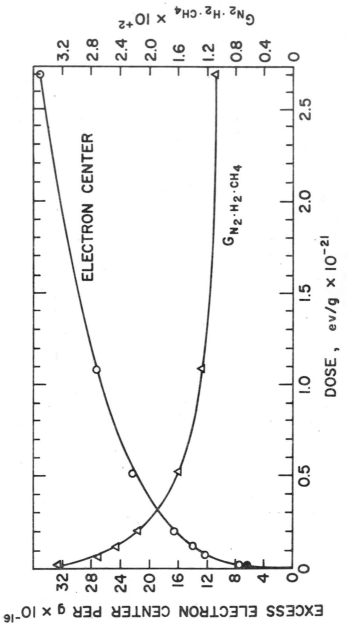

FIGURE 2-4   Decomposition of azoethane adsorbed on silica gel and production of *F* centres in irradiated NaCl

in the presence and absence of electron scavengers. The energy gap of ZrO_2 is about 5 eV and consequently this is the maximum energy which can be transferred. This amount of energy is not sufficient to decompose N_2O. If substantial decomposition is observed then presumably it would have to occur by dissociative electron capture. They observed decomposition in the N_2O–ZrO_2 system, however experiments with electron scavengers led them to attribute at least part of the decomposition of the N_2O to energy transfer via excited states of the solid (excitons). Ladov and Johnson[36] have concluded that energy transfer by excitons is a significant parameter in the decomposition of KNO_3. This fact is discussed later under the general heading "Nitrates". Rabe, Rabe and Allen[4] believe excited states of the solid are involved in their studies of azoethane absorbed an magnesium oxide. These authors found essentially zero energy transfer when substances with a relatively low band gap energy were used as the supporting material for the adsorbate; i.e. they found evidence for substantial energy transfer by insulators (MgO, SiO_2, alkali halides; band gap >7 eV) and essentially zero energy transfer with semiconductors (ZnO, TiO_2, NiO, graphite, band gap ~ 4 eV) Hentz *et al.*[32] provide evidence for rapid energy transfer in some silica gels, but also show that a significant fraction of the energy in some of these gels that is available for transfer to an adsorbate is trapped and is stable up to relatively high temperatures. Wong and Willard[37] have interpreted their results on the irradiation of adsorbed molecules on silica gel in terms of electron migration. For pure ionic solids, however, it is difficult to see how electron migration could be significant considering the Madelung field.

It is concluded that energy transfer is a significant parameter in the decomposition of inorganic solids, however all the available evidence indicates that this is only for the initial stages of the decomposition i.e. less than an absorbed dose of 1×10^{21} eV/g. Beyond this adsorbed dose it appears that energy transfer is not a significant parameter in the decomposition of these salts.

2.3 Back reaction

We will define the term "back reaction" to mean any radiation induced reaction of the decomposition products with each other or with the parent ion to produce products. For example the nitrates decompose to give nitrite ion and oxygen; back reaction in this system would be a reaction of oxygen with nitrite ion to produce nitrate ion.

Radiation induced solid state reactions have been known for a considerable time. Most of the early results however were obtained by those working in the general area of nuclear recoil reactions, and the actual characterization of solid state reactions, in the sense discussed here, was not sufficiently delineated to warrant inclusion here. The principal literature therefore on this subject is from the past two decades. The reactions will be discussed in detail under the appropriate headings (nitrates, chlorates, bromate, etc.) in what follows we will provide sufficient evidence to indicate the general occurence of back reactions in the systems of interest.

McCallum and Holmes[38] reported the presence of perchlorate ion in their studies on the $^{35}Cl(\gamma,n)^{34}Cl$ reaction in $NaClO_3$, this was subsequently confirmed by several others, however until recently the formation of ClO_4^- was thought to arise from the hydrolysis of Cl_2O_6 when the irradiated salt was dissolved in water. Evidence for the existence of ClO_4^- ion in irradiated $KClO_3$ by infrared absorption has been obtained by Hovi and Räsänen[39] and Brown and Boyd.[40]

Kinetic evidence for a back reaction was found in the decomposition of the alkali and alkaline earth perchlorates.[15] In this case because of the complexity of the reaction the back reaction was designated as

$$ClO_3^- + [products] \rightsquigarrow ClO_4^-. \qquad (2\text{-}1)$$

Evidence for a back reaction has been obtained in studies on the radiation induced decomposition of the alkali and alkaline earth bromates.[41] As with the perchlorates the evidence in these studies was kinetic i.e. kinetic analysis of the data. The back reaction postulated is that between energetic O atoms and bromite ion viz:

$$BrO_2^- + O \longrightarrow BrO_3^-. \qquad (2\text{-}2)$$

Recently Boyd and his associates have reported the presence of BrO_4^- ion in the radiation induced decomposition of the bromates.

Early studies on the decomposition of the nitrates gave results which indicated that the yield of product (NO_2^-) was linear with absorbed dose and consequently there could be no back reaction. It was subsequently shown[42] that back reaction between nitrite ion and the oxygen produced was indeed occuring. Recently Raman spectra[36] of KNO_3 enriched in oxygen-18 unequivically established the existence of a back reaction of the nitrite ion with oxygen. It may be added here that studies on the radiation induced decomposition of molten lithium nitrate also showed the existence of a radiation induced back reaction.[43]

Evidence for the occurence of back reaction has been obtained in studies on the decomposition of the sulfate hydrates. The sulfate ion is quite resistant to radiation induced decomposition however decomposition has been observed in the sulfates hydrates; in some of these hydrates only the water of hydration undergoes decomposition whereas others (nickel and iron sulfate hydrates) show some decomposition of the sulfate ion in addition to decomposition of the water of hydration. In all the compounds studied that exhibited decomposition evidence for back reaction was observed.[44]

No evidence appears to exist for back reaction occuring in the azides however the nature of the decomposition product (N_2) and the standard emf for the reaction

$$N_3^- \rightleftharpoons 3/2\,N_2 + e \quad E° = +3.1 \text{ V} \tag{2-3}$$

would indeed make a back reaction seem quite remote.

In summary therefore we may conclude that radiation induced back reaction is a general phenomena in the decomposition of inorganic ions and will occur if the reaction is energetically feasible.

2.4   Thermal decomposition

Many of the products of the radiation induced decomposition of inorganic molecular ions are not thermally stable and undergo thermal decomposition readily with changes in temperature. In some cases of course the products are quite stable (some are even more so than the parent ion). It is also found that the thermal stability of some of the product ions is different than that of a salt of the pure ion. For example ClO_2^- formed in the decomposition of the perchlorates is less stable (thermally) in the perchlorate lattice than in a pure chlorite lattice.[45] Annealing studies on radiation damage and temperature studies on radiation decomposition have revealed a considerable variety of interesting reactions which will be discussed in detail under the appropriate category of compounds.

Most radiation induced decompositions appear to have little or no temperature coefficient* i.e. there is no evidence that temperature influences the primary event. The effect of temperature, if any, appears to be related to secondary reactions of the primary products. For example ClO^- does not appear to be of primary product in the decomposition of alkali perchlorates.

---

* This does not apply to the radiation yield which may be affected due to changes in lattice parameters.

It arises from some complex radiation induced reaction of ClO_2 and ClO_2^- (products of the decomposition) but only above certain temperatures i.e. the reaction requires some thermal activation. In many cases, however, the primary products of the decomposition have not been delineated and consequently temperature effects have not been clearly resolved.

Several of the products of the decomposition of the perchlorates, chlorates and bromates undergo thermal decomposition with ease. In the perchlorates, chlorine dioxide, chlorite, and hypochlorite decompose (or react) slowly at room temperature.[15] The thermal stability of these compounds is different in the different lattices. For example the ClO^- ion appears to be more stable in the sodium perchlorate lattice than in the cesium perchlorate lattice. Annealing studies[45] in these compounds also revealed a solid state reaction occuring between an oxide and ClO_2 (a neutralization reaction).

In the bromates thermal decomposition and thermal reaction of the products occur.[46] The hypobromite ion, like the hypochlorite ion is thermally unstable and decomposes readily to give bromide ion and oxygen. Both this ion and the bromite ion appear to react on heating to reform bromate (the reaction occuring with oxygen in the lattice).

The decomposition products of the chlorates also exhibit some thermal decomposition as to be expected. Hypochlorite decomposes readily, however, some interesting thermally induced solid state reactions appear to occur. Annealing of irradiated $NaClO_3$ at $50°C$ produces an initial increase in the chloride, ion, however, continued heating at this temperature induces a reaction which produces a net loss in chloride ion,[47] presumably caused by the reaction of chloride ion with oxygen or an oxygen containing fragment. Irradiation of potassium chlorate at $25°C$ produces an absorption maxima at 4500 Å. This band is not present for crystals irradiated at $-196°C$ but does appear when these crystals are warmed to room temperature.[18]

High temperature annealing studies ($T \sim 120°$) on the nitrates indicates that the nitrite ion is quite stable toward thermal decomposition and only back reacts with oxygen to form nitrate. There are indications however that some interesting reactions occur at low temperatures. For example in potassium and silver nitrate it appears that NO_2 radicals form at $78°K$ but only if nitrite ion is present in the lattice i.e. irradiating pure KNO_3 at $78°K$ does not produce the NO_2 radical, the irradiated sample must be first warmed to room temperature (producing NO_2^-) and then subsequent irradiation at $78°K$ shows the presence of the NO_2 radical (this effect and others will be discussed in greater detail in a subsequent section).

No thermally induced reactions in the azides appear to occur however G values for azide ion decomposition increases with increasing temperature.[49] No significant differences in product yields or modes of decomposition in the sulfate hydrates were observed in the temperature range of 78–300°K.[44]

A radiation induced reaction of the products of the decomposition of KIO_3, I^- and O_2, to give IO_3^- and IO_4^- which appears to require some thermal activation has been observed.[158]

In summary it appears that the evidence for radiation induced secondary reactions requiring thermal activation is substantial and we may include that such reactions will occur when possible. In addition, of course, purely thermal solid state reactions occur and these reactions must be carefully separated from the radiation induced reactions referred to above.

References

14   G. Hennig, R. Lees, and M. Matheson, *J. Chem. Phys.* **21**, 664 (1953)

15   L. A. Prince and E. R. Johnson, *J. Phys. Chem.* **69**, 359 (1965), ibid **69**, 377 (1965)

16   G. E. Boyd, E. W. Graham, and Q. Larsen, *J. Phys. Chem.* **66**, 300 (1962)

17   J. W. Chase and G. E. Boyd ibid **70**, 1031 (1966)

18   C. E. Burchill, P. F. Patrick, and K. J. McCallum, *J. of Phys. Chem.* **71**, 4560 (1967)

19   E. R. Johnson, "Radiation-induced decomposition of inorganic solids" *Symposium on the Chemical and Physical Effects of High Energy Radiation on Inorganic Substances*, Atlantic City, New Jersey, June 1963, ASTM Special Publication No. 359

20   P. Doigan and T. W. Davis, *J. Phys. Chem.* **56**, 764 (1952)

21   I. Serlin, *Science*, **126**, 261 (1957)

22   H. W. Fenrick, S. U. Filseth, A. L. Hanson, and J. E. Willard, *J.A.C.S.* **85**, 3731 (1963)

23   F. W. Stacey, J. C. Sauer, and B. C. McKusick, *J.A.C.S.* **81**, 987 (1959)

24   J. W. Chase and G. E. Boyd, *J. Phys. Chem.*

25   P. W. Bridgeman, *Proc. Amer. Acad. Arts and Sci.* **51**, 581 (1916)

26   A. C. McClaren, *Rev. Pure and Appl. Chem.* Vol. 12, p. 54 (1962)

27   M. Balkanski, M. K. Teng, and M. Nusimouich, *Phys. Rev.* vol. 176 No. 3, 1098 (1968)

28   J. Cunningham, *J. Phys. Chem.* **65**, 628 (1961)

29   A. G. Maddock and S. R. Mohanty, *Sonderdruck* aus *Radiochemic Acta* 1, 85 (1963)
30   S. Okamura, K. Hayashi, and Y. Kitanishi, *J. Polym. Sci.* **58,** 927 (1962)
31   L. Kevan and W. F. Libby, *Recent Advances in Photochemistry*, 183 (1963)
32   See J. G. Rabe, B. Rabe and A. O. Allen, *J. Phys. Chem.* **70,** 1098 (1966) and E. A. Roja and R. R. Hentz ibid 2919 (1966)
33   A. R. Jones, *J. Chem. Phys.* **35,** 751 (1961)
34   M. Khare and E. R. Johnson, *J. Phys. Chem.* in press
35   N. H. Sagert and R. W. Robinson, *Can. Jour. of Chem.* **46,** 2075 (1968)
36   E. Ladov and E. R. Johnson, *J.A.C.S.* **91,** 7601 (1969)
37   P. K. Wong and J. E. Willard, *J. Phys. Chem.* **72,** 2623 (1968)
38   K. J. McCallum and O. G. Holmes, *Can. J. Chem.* **29,** 691 (1951)
39   V. Hovi and V. Räsänen, *Ann. Acad. Sci. Fenn.*, series A 6, 228 (1967)
40   L. C. Brown and G. E. Boyd, *J. Phys. Chem.* (1969)
41   J. W. Chase and G. E. Boyd, "Symposium on the effects of high-energy radiation on inorganic substances" Seattle, Wash. Oct. 31–Nov. 5, 1965 *ASTM Special Tech. Publ. No.* **100**
42   T. H. Chen and E. R. Johnson, *J. Phys. Chem.* **66,** 2249 (1962)
43   T. G. Ward, G. E. Boyd, and E. C. Axtmann, *Radiation Res.* **33,** No. 3, 447 (1968)
44   J. W. Chase and G. E. Boyd, "Symposium on the effects of high energy radiation on inorganic substances" Seattle, Wash. Oct. 31–Nov. 5, 1965 *ASTM Special Tech. Publ. No.* **100**
45   L. A. Prince, *Radiolysis of the Alkali and Alkaline Earth Perchlorates*, Thesis, Stevens Institute of Technology, Hoboken, New Jersey (1963)
46   G. E. Boyd and G. V. Larsen, *J. Phys. Chem.* **69,** 1413 (1965)
47   P. F. Patrick and K. J. McCallum, *Nature* **194**, No. 4830
48   H. G. Heal, *Can. J. of Chem.* **57,** 979 (1969)
49   H. G. Heal, *Trans. Far Soc.* **53,** 210 (1957)

3

Radiation induced decomposition of the inorganic nitrates

With the exception of the alkali halides, the nitrates have been the most extensively studied of any of the inorganic compounds to date. Recent reviews include that by Sviridov[50] who reviewed the existing literature up to 1962 and Zakharov and Nevostruev[51] who reviewed literature up to 1966. Much of the research reported on the decomposition of the nitrates is controversial and speculative leaving many unanswered questions. It appeared in the best interest of all to discuss in chronological order, all the research reported and when possible to indicate areas of difference. This manner of presentation may perhaps contain many redundancies, but it insures that undue emphasis will not be put on speculative concepts. In what follows, those individuals papers primarily concerned with the kinetics and mechanism of the decomposition will be discussed first. This will be followed by a discussion of optical absorptions bands, color centers, radicals and radicals ions and other general phenomenon. Following this there will be critical review of the existing data with a summary of "best values".

3.1   Products and yields

The stoichiometry of the room temperature decomposition of the alkali and alkaline earth nitrates may be described as

$$MNO_3 \rightsquigarrow MNO_2 + \tfrac{1}{2}O_2. \qquad (3\text{-}1)$$

Narayanswamy[52] studied the photochemical decomposition of a number of different nitrates using a quartz mercury arc and found that decomposition was restricted to surface layers. He believed that reaction in depth did

not occur because of back reaction of the nitrite with oxygen. Narayanswamy gave the following order in terms of increasing sensitivity to light as:
NH_4NO_3, $Pb(NO_3)_2$, $Al(NO_3)_3$, $Cd(NO_3)_2$, $Ba(NO_3)_2$, $Sr(NO_3)_2$,
$NaNO_3$, KNO_3. He also observed that light vibrating in the plane of the
nitrate ions was more effective than light vibrating normal to the planes
containing nitrate ions. Narayanswamy postulated the initial act to be

$$NO_3^- \longrightarrow NO_2^- + O \tag{3-2}$$

followed by $$NO_3^- + O \longrightarrow NO_2^- + O_2 \tag{3-3}$$

$$NO_2^- + O \longrightarrow NO_3^-. \tag{3-4}$$

Allen and Ghormley[53] studied the decomposition of barium nitrate with
fast electrons. They observed that the gas evolved upon dissolution of the
salt in water contained 3% hydrogen, the balance of the gas they assumed to
be oxygen. Solutions of the irradiated salt had an alkaline pH. They attributed the hydrogen gas and the alkaline reaction to be as a result of trapped
electrons. i.e. the trapped electrons reacting with the water as, $e^- + H_2O$
$\rightarrow \frac{1}{2}H_2 + OH^-$. These authors considered the primary act to be as indicated
in equation (3-2) but in addition also considered the possibility of the NO_3^-
ion losing an electron to form the NO_3 radical; the NO_3 decomposing to
give $NO_2 + O$ or attaching an electron to give $NO_2^- + O$. Although the
nature of the experiment did not provide accurate yields, the observation
was made that the conversion rate decreased with time.

Doigan and Davis[20] studied the photochemical decomposition of a large
number of nitrates. Nitrite ion was determined by Shinn's method however
the authors do not give the molar extinction at the wave length used (5400 Å).
They found the quantum yield to vary in terms of decreasing order as:
$CsNO_3 > La(NO_3)_3\,6\,H_2O > KNO_3 > Ba(NO_3)_2 > La(NO_3)_3 > Pb(NO_3)_2$
$> AgNO_3 > NH_4NO_3 > Sr(NO_3)_2 > NaNO_3 > LiNO_3$. The experiments of Doigan and Davis were done under more carefully controlled
conditions than Narayanswamy and hence the difference in the order of
sensitivity. Barium nitrate decomposition was linear with absorbed dose
up to 1% decomposition. Cesium nitrate was photolyzed to about 3%
decomposition. It was found that the rate of decomposition of this salt
decreased substantially after this amount of decomposition. The inference
was that the decomposition might be approaching a steady state. The effect
of temperature on the photolysis was studied, but only with the barium salt
which showed no effect in the temperature range of 20–250°C. These authors

also observed that water of hydration increases the sensitivity of the salts $La(NO_3)_3$ and $Cs(NO_3)_2$ toward decomposition.

Doigan and Davis suggested that since there appeared to be no effect of temperature in the photolysis that reactions (3-3) and (3-4) occured with equal activation energies or did not occur at all. These authors were led to the conclusion that the relative ease of decomposition of the nitrates is dependent on the field strength of the cation.

Hennig, Lees and Matheson[14] studied the decomposition of potassium and sodium nitrates induced by mixed pile irradiation and X-rays. They determined that the oxygen formed during the decomposition remains trapped in the lattice as oxygen molecules—not as atoms. This was established by magnetic susceptibility measurements in which they found agreement between theory and experiment to be 3.3% (within experimental error) and concluded that if any other paramagnetic substance were present it would have to present in very small amounts. The gas occluded in irradiated $NaNO_3$ was analyzed chemically for O_2, NO, NO_2 and H_2 and was found by these measurements to contain 98.5% oxygen, the remainder an inert gas presumably nitrogen. These findings were confirmed by mass spectrographic analysis. Annealing experiments revealed that on continued heating (annealing) of the irradiated crystals below the transition temperatures, that the trapped gas (oxygen) originally in small pockets, diffuses and coalesces to form larger pockets.

Hennig, Lees and Matheson found essentially no difference in G values between samples irradiated by fast neutrons (mixed pile irradiation) or gamma rays and concluded that atom displacement was not a significant factor in nitrate decomposition.* This was confirmed by experiments with X-rays whose energy was not sufficiently high to cause atom displacement. Although accurate G values for the decomposition were not available it was observed that KNO_3 underwent substantially more decomposition than $NaNO_3$ for the same absorbed dose. These authors postulated, that since there was no appreciable difference in the bond energies between the nitrate ion in the $NaNO_3$ and KNO_3 the differences they observed in 100 eV yields between the salts could best be attributed to the difference in free space between the salts (see Chapter II for discussion of this topic). These ex-

---

* D. Hall and G. N. Walton report in papers discussed later (reference 61) that energy deposition due to neutrons is only 1.4% that of gamma rays for samples irradiated in the moderator of a reactor core.

3*

periments led these authors to concur with previous workers[52,53] that the decomposition must proceed via ionization and or excited states of the nitrate ion. As with the results obtained by Allen and Ghormly these authors found a decrease in the G value at high absorbed doses.

Smith and Aten[54] studied the decomposition of $LiNO_3$, $NaNO_3$, $NaNO_2$, KNO_3 and NH_4NO_3 utilizing the $(n, 2n)$ reaction on nitrogen. They observed that all the N activity could be accounted for in NO_3^-, NO_2^- and nitrogen containing gases. The major constituent of the nitrogen containing gases was N_2 and the remainder appeared to be N_2O. Only 7% of the ^{13}N activity was in the gaseous products from the decomposition of $NaNO_3$ and $LiNO_3$ whereas 58% of the ^{13}N activity was in the gaseous products from $NaNO_2$ and NH_4NO_3 decomposition. No such data was available on KNO_3 decomposition because of the interference by ^{38}Cl (produced by nuclear reaction with potassium). These authors also observed that the yield of NO_2^- in $LiNO_3$ and $NaNO_3$ were comparable but appreciably less than in KNO_3.

Hochanadel and Davis[55] irradiated a series of nitrates at room temperature with cobalt-60 gamma rays and determined initial yields of nitrite ion. In agreement with others[52,53,14] they assumed the initial act to be dissociation of excited nitrate ions formed directly or formed by electron capture by the NO_3 radical to give $NO_2^- + O$. This primary act followed by a competition between the oxygen atom either recombining with the NO_2^- simultaneously formed or with an adjacent nitrate ion to form $NO_2^- + O_2$. They observed a linear production of NO_2^- with dose and concluded that there was little or no back reaction with oxygen atoms that escaped initial recombination. They determined G' values which are the G values divided by the ratio of electrons in the nitrate ion to the total number in the salt and compared relative G' values with the relative quantum yields from reference 20 and found reasonable agreement. They concluded from this that the energy imparted to the nitrate ion was more effective in causing decomposition than that to the cation. The results obtained by these authors are summarized in Table 3-1.

Cunningham and Heal[56] studied the decomposition of the alkali nitrates and the nitrates of barium, strontium and silver. The radiation source was X-rays with peak voltage of 44.5 kVp operated at a current of 45 milliamps. Dosimetry was determined calorimetrically.[57] They observed that all the salts except KNO_3 become light yellow to orange brown when irradiated at room temperature; potassium nitrate became deep orange when irradiated

TABLE 3-1 Nitrite yields from the gamma-ray decomposition of solid nitrates[a]

Salt	Mass abs. coeff. $\mu - \sigma_s \over \varrho$	NO_2^- ions per 100 eV		Relative yields		
		$G(NO_2^-)$	$G'(NO_2^-)$	G_{rel}	$G'_{rel'}$	θ_{rel}
$CsNO_3$	0.0246	1.68 ± 0.05	4.55	1.0	1.0	1.0
KNO_3	0.0263	1.57 ± 0.05	2.45	0.95	0.54	0.51
$NaNO_3$	0.0262	0.25 ± 0.02	0.35	0.15	0.07	0.06
$LiNO_3$	0.0262	0.02 ± 0.007	0.02	0.012	0.004	0.01
$AgNO_3$	0.0257	0.02 ± 0.05	0.50	0.12	0.1	0.19
$Ba(NO_3)_2$	0.0250	1.88 ± 0.15	3.68	1.12	0.81	0.25
$Pb(NO_3)_2$	0.0319	0.44 ± 0.04	1.05	0.26	0.23	0.19
$La(NO_3)_3$	0.0255	1.5 ± 0.2	2.50	0.91	0.55	0.22
$La(NO_3)_3\ 6\ H_2O$	0.0258	2.5 ± 0.3	5.6	1.52	1.23	0.63
Water	0.0294		3.2^n		0.71	

Data from C. J. Hochananadel and T. W. Davis *J. Chem. Phys.* **27**, (1957).
[a] The *G* values shown were calculated by using 15.6 molecules of Fe oxidized per 100 eV.

at $-195°C$ which turned to light yellow upon warming to room temperature. The color was stable but could be destroyed by subsequent irradiation.

Cunningham and Heal did a rather intensive study of the decomposition of KNO_3. A sample of the gas evolved from a salt irradiated to 32% decomposition was analyzed mass spectrographically and found to be over 99.5% oxygen. The gas contained no hydrogen, nitrogen or oxides of nitrogen. They observed that within experimental error the oxygen evolved was stoichiometric. Varying the intensity by a factor of 2.5 did not affect the decomposition. Furthermore these authors found the reaction to be first

TABLE 3-2 Initial *G* values for KNO_3 as a function of temperature

Temp. °C	Initial *G*	Temp. °C	Initial *G*
-186	$1.81 \pm .10$	85	$2.23 \pm .10$
-78	2.06	118	2.39
-40	2.04	128	2.56
-19	2.12	150	3.01
-11	2.04	173	3.13
15	1.96	200	3.34
52	2.16		

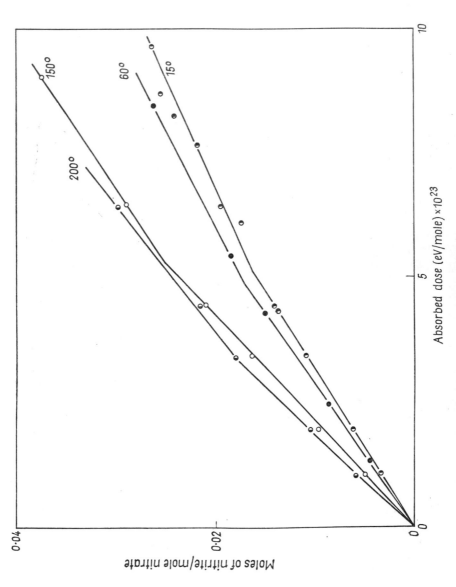

FIGURE 3-1   Decomposition of KNO_3 as a function of absorbed dose

order in the decomposition of nitrate except that there appeared to be a change in slope at absorbed dose of about 4–5 $\times$ 10^{21} eV/g (about 1.5 mol % decomposition). This apparent "break" in the yield versus dose curve amounted to two intersecting linear curves indicating first order reaction in each portion of the decomposition; the rate after the "break" being only 70% that prior to the "break" (see Figure 3-1).

The effect of previous history of the crystals was also examined. Lithium nitrate in two different anhydrous forms showed a remarkable difference in initial G values (see Table 3-3). However the same G value was found for potassium nitrate which had respectively been fused and quenched, fused and cooled slowly or recrystalized from water. Initial G values for the different nitrates obtained in this study are summarized in Table 3-3.

TABLE 3-3   Initial G values for the decomposition of the nitrates at different temperatures

Salt	T°C	G
$LiNO_3$ (fused)	15	0.02 $\pm$ .01
(dehydrated cold)	15	0.006 $\pm$ .003
$NaNO_3$	15	0.37
	150	1.10
$RbNO_3$	15	0.64
	150	1.37
$CsNO_3$	15	1.37
	150	1.17
$AgNO_3$	15	0.14
$Ba(NO_3)_2$	15	1.75
	150	1.84
$Sr(NO_3)_2$	15	0.52
	150	1.33

X-ray diffraction and IR (infrared) measurements on samples of KNO_3 irradiated to different percentages decomposition indicated the presence of the nitrite ion in the KNO_3 lattice. IR studies were also obtained on the other nitrates and these like KNO_3, revealed the presence of NO_2^- in the nitrate lattice. Although the presence of nitrite ion was apparent after 4 % decomposition in the IR studies, it was not until decomposition was greater than 8% that the presence of nitrite ion could be detected by X-ray diffraction (the pattern due to the presence of nitrite ion was not observed in a sample that had undergone 8% decomposition). Cunningham and Heal concluded from their X-ray studies that since the nitrite ion line was sufficiently intense so that it could have easily been observed if it had only one-fifth the intensity found, (for a sample decomposed to 8%) decomposition must have occurred at random isolated sites throughout the lattice not in clusters at some preferred site or sites.

The observed differences in G values of the various nitrates was related to the "free space". They obtained an excellent fit of their data for plots of the G values for alkali nitrates versus free space, however here as with other studies discussed in Chapter 2, agreement was not good if the alkaline earth nitrates were included in the plot.

The dosimetry employed by Cunningham and Heal was calormetric and apparently in error since subsequent studies by others (including Cunningham) indicate that the G values reported are too high. In addition these authors assumed that all the nitrates decompositions were linear with absorbed dose which is also in error.

The mechanism postulated for the decomposition was in agreement to that of others i.e. the initial act was considered to be ionization or excitation of the nitrate ion. The NO_3 radical formed in the ionization act could dissociate to give an NO_2 radical plus oxygen, the NO_2 radical subsequently combining with an electron to form nitrite ion, or the NO_3 radical could undergo a dissociative attachment reaction with an electron to give nitrite ion plus an oxygen atom. The excited nitrate ions could return to the ground state and give up their energy to the lattice or dissociate to nitrite ions plus oxygen. The apparent break in the NO_2^- yield versus dose curve was explained by assuming that the oxygen atoms initially formed are trapped at lattice vacancies. When the number of oxygen atoms exceeds twice the number of traps available oxygen molecules form in normal lattice regions in an amount equivalent to the solubility of oxygen in the nitrate lattice. These "dissolved" oxygen molecules compress the nitrate molecules in their vicinity and hence cause a reduction in the "free space". This reduction in the free space accounting for the change in slope observed in KNO_3 at about 1.5 mol-% decomposition.

Johnson[58] studied the decomposition of lead nitrate and observed the reaction to be complex. Nitrite yields appeared to be linear with dose however oxygen was found to be initially in a 1 : 1 ratio with nitrite but as the decomposition proceeded it, fell to the stoichiometric ratio. These results are summarized in Figure 3-2. To account for this change in oxygen

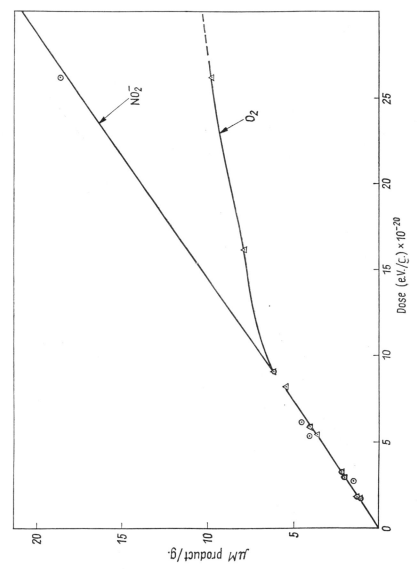

FIGURE 3-2   Nitrite and oxygen production from the decomposition of $Pb(NO_3)_2$

yields Johnson postulated the following sequence of reactions:

$$Pb(NO_3)_2 \xrightarrow{\sim\!\!\sim\!\!\sim} Pb + 2\,NO_2 + O_2 \tag{3-5}$$

$$2\,NO_2 + H_2O \longrightarrow 2\,H^+ + NO_2^- + NO_3^- \tag{3-6}$$

$$Pb + \tfrac{1}{2}\,O_2 \xrightarrow{\sim\!\!\sim\!\!\sim} PbO. \tag{3-7}$$

Solutions of irradiated nitrate were found to have an amber color with no absorption in the visible or UV spectrum and the color was attributed to a colloidal suspension of PbO. Reflectence spectra of irradiated. $Pb(NO_3)_2$ showed the presence of an absorption maxima at about 3500 Å which was attributed to metastable nitrate ions.* A G values of about 0.43 was obtained for the decomposition of this salt and a G value of 1.38 for KNO_3.

Forten and Johnson[59] examined the product yield and heats of solution of irradiated KNO_3. Heats of solution measurements were done by twin calorimetry. The source of radiation was a cobalt-60 and ferric sulfate dosimetry was used with $G(Fe^{3+}) = 15.5$. They observed from heats of solution measurements that at about 1 mole % decomposition the KNO_3 lattice appeared to undergo a low energy transition (see Figure 3-3). Density measurements on the irradiated salt indicated a sharp decrease in density of about 1% at an absorbed dose just beyond that producing the transition. A plot of nitrite yield versus absorbed dose appeared to be linear and like that observed by Cunningham and Heal could be separated into two intersecting straight lines. The apparent "break" in the yield versus dose curve appeared to occur at the same point observed by Cunningham and Heal. The change in density and the transition observed in the heats of solution measurements were coincident with the "break" in the nitrite yield versus dose curve. The experimental error, however, in the results obtained by Johnson and Forten was sufficiently great so that the apparent "break" was not clear cut. Cunningham and Heal observed a band at 1250 cm^{-1} in the IR spectra of irradiated KNO_3 which was attributed to isolated nitrite ions modified by dispersing in the KBr lattice (KBr pelleting techniques). Forten and Johnson also observed this band but attributed it to a KNO_3 lattice modified by the transition they believed to have occurred.

A number of bivalent nitrates were studied by Baberken[60] with the purpose of examining the role of the cations in the decomposition of these

---

* It is now known that NO_2 is indeed a primary product in the decomposition of $Pb(NO_3)_2$ and hence the band at 3400 Å could well be NO_2 trapped at lattice sites.

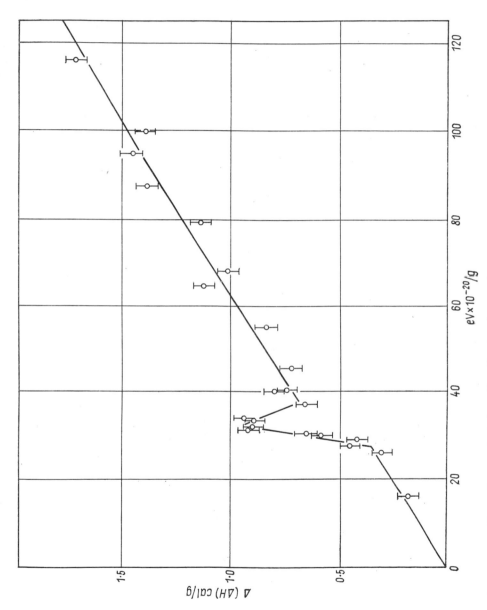

FIGURE 3-3 Change in heat of solution of irradiated KNO_3 as a function of absorbed dose

salts. The radiation source was cobalt-60 and all irradiations were done at 20°C. Details of the dosimetry used in these experiments were not revealed. Baberkin calculated a mean G value of the nitrite formed and compared this with "free space", polarizability, and lattice energy. A correlation between "free space" and nitrite yield in the alkaline earth nitrates was obtained and one between cadmium and zinc nitrates, but the correlation did not hold if all the divalent nitrates were considered. Baberkin used a volume for the $NO_3^- = 31.6$ Å^3 which is considerably higher than the value 14.5 Å^3 used by Cunningham and Heal. The results obtained by Baberkin are summarized in Table 3-4 below.

The effect of fission product recoils on the decomposition of KNO_3 was studied by Hall and Walton.[61] The experiments were done by sandwiching pressed pellets of dry KNO_3 (0.5 in. diam.) between disks of uranium foil. The sandwiched material was held in place by a steel clip and the entire sandwich sealed into an evacuated glass capsule and exposed to mixed pile radiation which had a neutron flux of 1.4×10^{12} neutrons per cm^2 per sec. After several days exposure the KNO_3 was dissolved and the nitrite determined by Shinn's method. The relative neutron doses were determined by adding cobalt wire to each sample and determining the cobalt-60 activity. The concentration of fission recoils was determined by analyzing for ^{89}Sr and ^{140}Ba. The decomposition of pure KNO_3 by pile irradiation was determined in separate experiments and the decomposition induced by the recoils corrected for accordingly. Hall and Walton estimated that the number of atom displacements per fission fragment was about 8.6×10^4; the observed value for the number of nitrate ions decomposed per fission fragment was 8.6×10^6. From this difference the authors concluded that atom displacement is not the primary cause of decomposition. The track of a fission fragment was estimated to be a cylinder 1.76×10^{-3} cm long with a diameter of about 100 Å. The energy absorbed in this track correspond to a temperature of about 4000°C; the duration time being about 10^{-11} sec. This temperature is sufficient to destroy all lattice bonding along the track. From the data obtained by the authors the total number of molecules decomposed per fission fragment was 8.2×10^6 from which they estimated, using the known value for the density of KNO_3, that this number of molecules would occupy a cylinder 1.76×10^{-3} cm long with a diameter of 70 Å approximately the size of the thermal spike calculated for the fission recoil. The G value obtained in these studies was found to be 10.5 The authors believe that this difference in G values between recoils and X- or gamma-radiation

TABLE 3-4 Correlation of G values of some divalent nitrates with various parameters

Cation	Charge	Radius Å	No. of electrons in the outer shell	Free volume of the unit cell per NO_3^- ion, Å³	Free volume of the unit cell per NO_3^- ion, Å³ (according to calc.)	Polarisability	Lattice energy, kcal/mole	Electron contribution of NO_3^-	Mean quantity of nitrite produced in molecules per 100 eV
Ba^{2+}	2	1.43	8	47.0	29.0	1.86	457.47	0.54	0.33
Sr^{2+}	2	1.27	8	40.4	24.3	1.02	464.72	0.64	0.14
Ca^{2+}	2	1.04	8	72.2	20.1	0.552	505.47	0.78	0.10
Cd^{2+}	2	0.99	18	—	37.0[a]	—	522.12	0.57	0.16
Cd^{2+}	2	0.99	18	—	74.0[a]	—	522.12	0.57	0.16
Zn^{2+}	2	0.83	18	—	32.2³	—	568.32	0.70	0.13
Zn^{2+}	2	0.83	18	—	66.4[a]	—	—	0.70	0.13

[a] Calculated from crystal-chemical electronegativities.

is due to the general destruction of the lattice and the ability of the oxygen atoms formed in the initial act to escape recapture.

In a subsequent paper, Hall and Walton[62] repeated their experiments using thin films of uranium oxide enriched in ^{235}U. They recognized that their previous calculations on the energy absorbed by the KNO_3 were in error. This error was in part due to the thickness of the uranium foil used, and the fact that decomposition of the KNO_3 by pile background radiation was comparable to the decomposition by the fission fragments. The uranium foils had a thickness in excess of the range of the recoils. The radiation (gamma, beta, etc.) from the fission process had a significant effect on the decomposition. Very careful experiments using aluminium absorbers, different film thickness, range experiments etc. permitted a better evaluation of the nitrite yield as a function of energy absorbed. The value obtained in this study was 6.0 molecules of NO_2^- formed per 100 eV absorbed.

Cunningham[63] in a continuation of his earlier studies on the radiation induced decomposition of nitrates, made an extensive study of the effect of temperature on the decomposition of the alkali nitrates and barium nitrate. In addition he studied the isotope effect in the decomposition of potassium nitrate. As in the previous study Cunningham observed that the decomposition was linear with dose except that a sharp "break" in the curve was observed (as before) at an absorbed dose of about 5×10^{21} eV/g. He observed essentially zero isotope effect for the nitrogen isotopes (^{15}N vs ^{14}N) but a 12% isotope effect between $KN^{18}O_3$ and $KN^{16}O_3$, however only in the initial region of yield vs. dose curve (prior to the "break") Figure 3-4. Beyond this point he found essentially zero isotope effect. Furthermore Cunningham found essentially zero isotope effect in the initial decomposition at $-110°C$, and $81°C$; at $122°$ the isotope effect was about 5% but at $190°C$ the isotope effect reversed i.e. the decomposition of the oxygen-18 containing salt was greater than that of the oxygen-16 containing salt. This was also true in the liquid state. These results are summarized in Tables 3-5 and 3-6.

TABLE 3-5 $G(NO_2^-)$ for Isotopically substituted salts

Salt	Initial $G(NO_2^-)$	After the observed "break" $G(NO_2^-)$
$K^{14}N^{16}O_3$	1.456 ± 0.011	0.902 ± 0.029
$K^{14}N^{18}O_3$	1.298 ± 0.017	0.888 ± 0.037
$K^{15}N^{16}O_3$	$1.435 + 0.019$	. . . .

TABLE 3-6 Comparison of initial $G(NO_2^-)$ for the various nitrates at different temperature

Temp. °C	$NaNO_3$	$KN^{16}O_3$	$KN^{18}O_3$	$RbNO_3$	$CsNO_3$	$Ba(NO_3)_2$	$LiNO_3$
−110	0.341 ± 0.013	1.472 ± 0.017	1.428 ± 0.023	0.79 ± 0.05	1.346 ± 0.043	1.426 ± 0.043	—
25	0.200 ± 0.004	1.456 ± 0.011	1.298 ± 0.017	0.60 ± 0.1	1.72 ± 0.10	1.80 ± 0.10	—
60	0.308 ± 0.02	1.46 ± 0.04	1.40 ± 0.04	0.78 ± 0.05	1.66 ± 0.50	1.78 ± 0.06	—
81	0.47 ± 0.03	1.62 ± 0.03	1.40 ± 0.04	0.95 ± 0.05	1.62 ± 0.06	1.73 ± 0.06	—
122	0.67 ± 0.02	2.25 ± 0.03	2.13 ± 0.04	1.11 ± 0.05	2.12 ± 0.08	1.89 ± 0.06	—
190	1.04 ± 0.10	2.84 ± 0.20	2.98 ± 0.20	2.54 ± 0.10	3.13 ± 0.15	1.81 ± 0.08	—
330	5.1 (L)	5.8 (L)	9.9 (L)	4.5 (L)	7.0 (L)	0.8 (L)	0.07 (L)

The principal isotope effect observed by Cunningham was only at 25°C and this occurred only in the initial portion of the decomposition. As can be seen the initial $G(NO_2^-)$ for normal KNO_3 is constant with temperature up to 60°C, beyond this temperature initial $G(NO_2^-)$ increases.

Cunningham developed a diffusion controlled mechanism and explained the large isotope effect observed at 25°C by the Franck–Rabinowitch cage effect* and the jump-frequency of the oxygen fragment produced in the initial act (equation 3-2). He indicated that for those nitrates with the smallest free space, $G(NO_2^-)$ is greater at low temperatures $(-110°C)$ than at higher temperatures, while for the nitrates with higher free space $G(NO_2^-)$ at 25° is higher than at the lower temperatures. From these observations he concluded that deactivation of excited species by lattice collision were responsible for

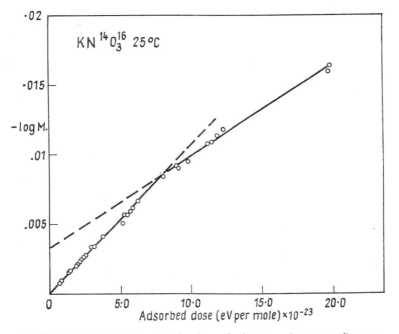

FIGURE 3-4   Plot of log M (fraction of nitrate undecomposed) as a function of absorbed dose

* The cage effect may be viewed as follows: the oxygen atom released in the initial act, equation 3–2, will find itself surrounded by the atoms (primarily oxygen) of nitrate ions which may be visualized as a cage. In order for the oxygen atom to escape it must of necessity undergo collision with the atoms of the cage. If these "cage" atoms are heavy the likelyhood of the oxygen atom escaping the cage are less than if they are light atoms.

the decrease in G from $-110°$ to $25°$. At about $25°$ sufficient expansion of the lattice occurs so that the jump distance of the oxygen fragment should increase leading to an increase in $G(NO_2^-)$. This fact and the operation of the Franck–Rabinowitch cage effect account then for the large isotope effect at $25°$. If this supposition is correct then it follows that the G value for the normal salt should show an increase with temperature which it does not. In addition, the mechanism proposed is unable to account for the decrease in the isotope effect beyond an absorbed dose of 6×10^{21} eV/g.

Further studies on heats of solution measurements of the irradiated nitrates were reported by Johnson and Forten.[64] Heats of solution curves for $CsNO_3$ and $NaNO_3$ and the NO_2^- yield versus dose curves were obtained for these salts plus that for $Pb(NO_3)_2$. The heats of solution curves for $CsNO_3$ and $NaNO_3$ like KNO_3 exhibited maxima; that for $CsNO_3$ had two maxima while that for $NaNO_3$ exhibited what appears as several abrupt increases in the heat of solution with absorbed dose. The authors concluded that the maxima in the heats of solution were caused by the strain induced in the lattice by the products, notably oxygen. Subsequent irradiation causes some expansion of the lattice permitting gas molecules to diffuse and coalesce into pockets, relieving the strain. The largest increase in the heats of solution measurements was found with $NaNO_3$. This lattice had the closest packed structure of all the nitrates studied.

The nitrite yield versus dose curve was plotted as a smooth curve for $NaNO_3$ decomposition. A kinetic expression for nitrite yield versus absorbed dose was derived which appeared to give excellent agreement with experiment. This was:

$$NO_3^- \rightsquigarrow NO_2^- + O \qquad k_1\phi \qquad (3\text{-}8)$$

$$O + NO_2^- \longrightarrow NO_3^- \qquad k_2 \qquad (3\text{-}9)$$

$$O + NO_3^- \longrightarrow NO_2^- + O_2 \qquad k_3 \qquad (3\text{-}10)$$

This is essentially the reaction scheme originally suggested by Narayanswamy. Application of the steady state assumption for oxygen atoms gives:

$$\frac{d(NO_2^-)}{dt} = \frac{2k_1k_3\phi(NO_3^-)^2}{k_2(NO_2^-) + k_3(NO_3^-)} \qquad (3\text{-}11)$$

which on integration gives:

$$\frac{k_2}{k_3} 2.303 \log (1 - X) + \frac{k_2/k_3 X}{1 - X} - 2.303 \log (1 - X) = 2k_1\phi t$$

where $X = (NO_2^-)/[NO_3^-]_0$. \qquad (3\text{-}12)

NO_2^- = concentration of (NO_2^-) at time t, $[NO_3^-]_0$ = initial nitrate concentration.

It was suggested by the authors that such an expression would also be valid for nitrate decompositions other than $NaNO_3$.

Jones and Durfee[65] irradiated KNO_3 dispersed in pressed KBr pellets similar to those used in infrared analysis. The disappearance of nitrate ion and the appearance of nitrite ion were analyzed for by infrared using the $1385\ cm^{-1}$ band due to NO_3^- and the $1270\ cm^{-1}$ band due to NO_2^-. A quantitative estimate of nitrite in using IR technique was developed. The NO_2^- was also analyzed for by Shinn's method (molar extinction coefficient at 5350 Å was 43,100). The pellets were prepared by mixing a slurry of KBr with a dilute KNO_3 solution and then drying with simultaneous grinding. The dried powders were further dessicated by drying in a vacuum oven at 115°C. This technique produced essentially isolated nitrate ions in the bromide lattice (Jones[66] had determined that the solubility of KNO_3 in KBr was about 0.21 mole-%). One of the principal reasons for doing this particular experiment was to investigate the possibility of formation of free radicals or radical ions that might be stable in KBr matrix. Infrared examination of the irradiated pellets did not reveal any such species. However the studies produced some very interesting results. For one, excellent quantitative agreement was found between the colormetric method of determining NO_2^- and the infrared method. This unequivically established that the nitrite ion, found by the usual colorimetric analysis, existed as such in the solid prior to dissolution and quantitative estimation. Nitrite decomposition was not linear in these systems; the disappearance of NO_3^- versus dose leveled off as did the nitrite yield. This appeared to indicate the presence of back reaction of oxygen with nitrite ion. When KNO_2 only was dispersed in the KBr lattice and then irradiated, decomposition of the nitrite ion was observed which was accompanied by a small amount of nitrate formation. These results appeared to indicate that nitrite ion decomposition did not yield substantial amounts of oxygen (or oxygen atoms) which could react with nitrite to form nitrate ion. Unfortunately Jones and Durfee do not give any absolute dosimetry and consequently it is not possible to compare these results except qualitatively with those on the pure salt.

Johnson[67] studied the effect of intensity on the decomposition of KNO_3, $NaNO_3$, $CsNO_3$ and $Pb(NO_3)_2$. The crystals were irradiated with 1.5 MeV electrons from a Van de Graaf accelerator. Dosimetry was determined by multiplying the total charge collected $\times$ time $\times$ voltage. Dose rates varied for KNO_3 from 10^{17} eV/g-sec to 10^{20} eV/g-sec; for $CsNO_3$ and $NaNO_3$ the dose rates were 10^{19}–10^{20} eV/g-sec. No variation in yield ($G(NO_3^-)$) was found over the entire dose range studied. Good agreement between these studies and those using gamma rays (dose rate $= 10^{15}$ eV/g-sec) was obtained. The initial yield for KNO_3 decomposition was found to be 1.48 ± 0.21, for $NaNO_3$ $G = 0.22 \pm .02$, for $CsNO_3$ $G = 1.53 \pm .08$ and for $Pb(NO_3)_2$ $G = 0.53 \pm .08$. These experiments established that since no significant difference in initial yields were observed when the dose rate was varied from 10^{15} eV/g-sec to 10^{20} eV/g-sec that oxygen atoms (or other active species free to migrate) did not enter into competition reactions such as:

$$O + O \longrightarrow O_2 \qquad (3\text{-}13)$$

$$O + [NO_3^- \text{ or } NO_2^-] \longrightarrow [NO_2^- + O_2 \text{ or } NO_3^-] \qquad (3\text{-}14)$$

i.e. reaction (3-13) is not an important competitive reaction in the decomposition of the nitrates.

Hochanadel[68] irradiated a number of nitrates with ^{210}Po alphas to ascertain the effect of LET (linear energy transfer). The alpha source consisted of 2 curies of polonium-210 plated on a flat stainless steel disk. All samples were irradiated as powders. Alpha particle dosimetry was done by irradiating rapidly stirred solutions of ferrous and ceric sulfate. The dosimetry solutions were analyzed continuously by spectrophotometry. Hochanadel determined that the G value for ferrous sulfate oxidation in his system was 3.14 and for ceric reduction 2.87. Irradiations at 30 and 150°C were also performed using cobalt-60 gamma rays; ferrous sulfate dosimetry was used with $G(Fe^{3+}) = 15.6$. In calculating initial G values Hochanadel used the true mass absorption coefficients shown in Table 3-1. For comparison with the gamma and alpha radiolysis photochemical decompositions were also studied using a Hanovia low pressure mercury discharge lamp. In the photolytic experiments the rate of energy absorption was determined by assuming a quantum yield of 0.1 for the decomposition of KNO_3 at 40°C. The results obtained by Hochanadel are summarized in Table 3-7.

Hochanadel observed that the increased yield with increased LET roughly parallels the increase in yield with temperature for gamma-radiolysis and suggested that the substantial increase in decomposition yield for $NaNO_3$ by alpha particles could be attributed to local heating. Hochanadel, however, did not rule out the possibility that the effect of high LET could also be

4*

explained on the spacial distribution of intermediates (competition reactions of intermediate species).

Chen and Johnson[42] studied the room temperature decomposition of several nitrates to determine if the kinetic mechanism found for $NaNO_3$ was applicable to other nitrates. The radiation source was a fixed cylinder of cobalt-60 located at the base of an 8 ft. shaft ("hole in the ground" type of installation) which permitted very reproducible positioning of the samples in the source. Ferrous sulfate dosimetry was used with $G(Fe^{3+}) = 15.45$.

TABLE 3-7   Initial G values for nitrite formation from various nitrates using different energy sources

Salt	Radiation		Ultraviolet light
	3.4 MeV α-rays	γ-rays	
$NaNO_3$	1.3 (25°)	0.27 (30°)	0.2 (40°)
	1.1 (120°)	1.0 (150°)	0.6 (150°)
KNO_3	2.2 (25°)	1.5 (30°)	1.6 (40°)
		3.0 (150°)	2.5 (150°)
$CsNO_3$	1.4 (25°)	1.6 (30°)	
		2.3 (150°)	
$Ba(NO_3)_2$	1.6 (25°)	1.8 (30°)	
		1.6 (150°)	
$LiNO_3$	0.7 (25°)	0.02–0.2 (30°)	
		0.03–0.4 (150°)	

Absorbed doses were calculated using the true mass absorption coefficients given in references 55 however, the value of 0.025 (instead of 0.0308) was erroneously used for the mass absorption coefficient for the dosimetry solution. The reported G values must be corrected accordingly. It was found that the kinetic equation expressing the formation of nitrite ion with absorbed dose [equation (3-11)] was also applicable to KNO_3, $CsNO_3$, $Ba(NO_3)_2$, and $Pb(NO_3)_2$ decomposition but not $AgNO_3$. For KNO_3 decomposition the authors observed that although the yield was not linear with dose the small variation in G values was such that if there was an overall error of $\pm 5\%$ the curve could be redrawn as 2 or 3 straight lines. They attributed the apparent "break", therefore, observed by Cunningham and Heal,[56] Forten and Johnson,[59] and Cunningham[63] as due primarily to this fact. Furthermore, they showed that at low doses, equation (3-11) could be simplified to give a linear expression

$$X = bT \qquad (3\text{-}15)$$

where X = nitrite formed
T = absorbed dose
b = constant.

These results adequately explained the first order dependence observed by previous workers. Equation (3-11) may also be integrated assuming that for small amounts of decomposition the NO_3^- concentration remains essentially constant. This gives

$$\frac{k_2}{2k_3} \frac{1}{(NO_3^-)} (NO_2^-)^2 + (NO_2^-) = 2k_1\phi(NO_3^-)\,t. \qquad (3\text{-}16)$$

For KNO_3 decomposition equation (3-12) was used with the values of the constants obtained from a least squares fit of the experimental data. This gave the following expression for nitrite versus absorbed dose

$$13.3 \left\{ 2.303 \log [1 - X] + \frac{X}{1 - X} \right\} - 2.303 \log (1 - X) = 0.00181T$$

$$(3\text{-}17)$$

where $X = (NO_2^-)/(NO_3^-)_0$
NO_2^- = nitrite ion concentration at time t
$(NO_3^-)_0$ = initial nitrate ion concentration
T = absorbed dose.

For $CsNO_3$, $NaNO_3$, $Ba(NO_3)_2$, and $Pb(NO_3)_2$ decomposition equation (3-16) was used with the values of the constants obtained from a least squares fit of the experimental data.

$$CsNO_3: \quad 0.66\,(NO_2^-)^2 + (NO_2^-) = 3.53\,T \qquad (3\text{-}18)$$

$$NaNO_3: \quad 0.182\,(NO_2^-)^2 + (NO_2^-) = 1.49\,T \qquad (3\text{-}19)$$

$$Ba(NO_3)_2: \quad 0.0735\,(NO_2^-)^2 + (NO_2^-) = 1.99\,T \qquad (3\text{-}20)$$

$$Pb(NO_3)_2: \quad 0.331\,(NO_2^-)^2 + (NO_2^-) = 0.599\,T. \qquad (3\text{-}21)$$

These equations give excellent agreement with experiments for the room temperature decomposition up to at least 3–4 mole-% decomposition.

Chen and Johnson[69] studied the effect of pressure on the decomposition of potassium nitrate. Samples of KNO_3 contained in a 2 mm ID pyrex tube were placed inside a stainless steel tube which was connected directly to a helium cylinder. The irradiation source was cobalt-60 and all pressure experiments were done under 1900 p.s.i. of helium. Experiments were done in the presence and absence of helium and it was determined that the high pressure helium absorbed less than 0.3% of the energy. Ferrous sulfate dosimetry was used assuming $G(Fe^{3+}) = 15.45$. The results indicated that there was no effect of pressure up to an absorbed dose of about 0.45×10^{21} eV/g but beyond this the authors observed a significant decrease in nitrite yield with absorbed dose. These results are shown in Figure 3-5.

Cunningham and Steele[70] studied the effect of pressure on the decomposition of KNO_3, $CsNO_3$, $NaNO_3$ and $RbNO_3$. All samples were contained in platinum and were then fitted into sample holders which could be sealed off under pressure. The radiation source was cobalt-60 and the pressure fluid, helium or nitrogen. All samples were sealed under 700 atmospheres of gas pressure. Absorbed doses were calculated by assuming that $G(NO_2^-)$ for unpressurized KNO_3 was 1.456. For the initial nitrite yield (up to an absorbed dose of about 0.6×10^{21} eV/g) the authors report that irradiation under 700 atmospheres pressure causes a decrease of 4.5% in the nitrite yield in KNO_3, 9.0% in $NaNO_3$, 19.6% in $CsNO_3$ and 22.3% in $RbNO_3$. The pressure used in these studies was considerably greater than that used by Chen and Johnson consequently the small decrease in yield observed in KNO_3 may possibly be real.

Logan and Moore[71] studied the decomposition of KNO_3, $LiNO_3$, $NaNO_3$, $RbNO_3$, and $CsNO_3$ using light ions (D^+, He^+ and D_3^+) at kilovolt energies. In the energy range considered, it is assumed that the incoming ion readily captures an electron and the subsequent interaction may be treated as a hard sphere collision. In effect they viewed their system as the incoming ion being treated as an extremely hot atom. The radiation source was a radiofrequency type operated from a 18.5 Mc oscillator. The ions were extracted from the source and accelerated to various energies. The samples to be irradiated were in the form of micro-crystalline coatings spead over a stainless steel block. The total amount of nitrate spead on the block (area = 5 cm²) was about 100 mg. Ion currents were determined by a coulometer.

The total yield (of NO_2^-) with absorbed dose showed a rapid increase followed by a leveling off with the formation of a plateau i.e. the yield saturated very quickly. As to be expected, the saturation yield was a function of bombardment energy (depth of penetration of the beam). They also

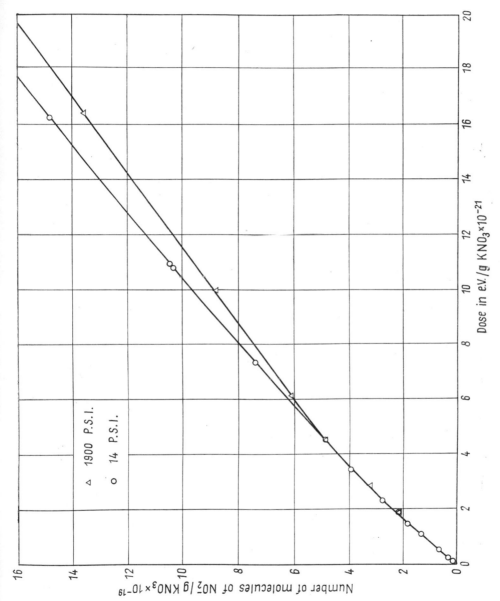

FIGURE 3-5

observed that decomposition of nitrite ion was neglible. Initial yields of nitrite formation are shown in Table 3-8.

The variation of initial G values for KNO_3 with temperature using 2 keV D^+ ions produced an interesting effect. The curve (Figure 3-6) has a distinct minimum at about room temperature; below and above this temperature there is an increase in yield.

Logan and Moore used the method developed by Hsiung and Gordus[71a] for the calculation of the minimum net recoil energy required to effect bond rupture following a collision, with momentum transfer to a polytomic

TABLE 3-8   Comparison of initial $G(NO_2^-)$ using 2 KeV D^+ ions

	$G(NO_2^-)$
$LiNO_3$	0.60
$NaNO_3$	1.80
KNO_3	2.00
$RbNO_3$	1.50
$CsNO_3$	1.40

molecule. Using this model the calculated G values for He^+ ions on KNO_3 gave good agreement with experiment while those for D^+ ions gave results consistently high. The model could not account for the difference in yields of the various nitrates. The authors explained the difference in yields as being due to the relative ease of transferring energy to lattice vibrational modes. Thus, they explained the low yield in $LiNO_3$ as being due to the tightness of the packing which raises the force content and the vibration frequencies; the density of vibration states and their higher average energies facilitate the dissapition of energy. A comparison of these results with those of reference 68 indicate a reasonably good correlation (except for $NaNO_3$), however, the correlation with the data of reference 63 on the effect of temperature on the gamma radiolysis shows some serious differences. These differences and the fact that a minimum is observed in the temperature vs. yield curve appear to indicate that the effect of LET cannot be explained by temperature effects alone.

Cunningham reported on the presence of several intermediates (as detected by ESR and optical absorption) which are produced when KNO_3 and $NaNO_3$ are irradiated at low temperatures. These results will be discussed in a later section. In a subsequent paper Cunningham[72] has reported on the chemical means of detecting these intermediates. The experimental technique for the determination of products (NO_2^-) were essentially similar to those reported previously (Shinn's method for NO_2^- analysis) except that

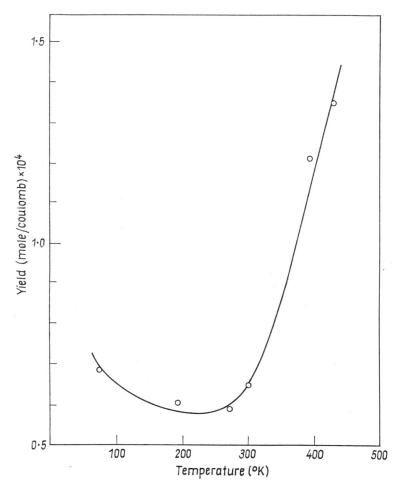

FIGURE 3-6   Variation in initial G values with temperature for the decomposition of KNO_3 with 2 keV D^+ ions

concurrent analyses for products were also done using dilute ceric sulfate in 0.4M H_2SO_4 Cunningham had determined that the reaction

$$NO_2^- + 2\,Ce^{4+} + H_2O \longrightarrow NO_3^- + 2\,Ce^{3+} + 2\,H^+ \qquad (3\text{-}22)$$

was quantitative and that the reactions

$$2\,H_2O + 2\,NO_2 + 2\,Ce^{4+} \longrightarrow 3\,NO_3^- + 2\,Ce^{3+} + 4\,H^+ \qquad (3\text{-}23)$$

and

$$NO + 3\,Ce^{4+} + 2\,H_2O \longrightarrow NO_3^- + 3\,Ce^{3+} + 4\,H^+ \qquad (3\text{-}24)$$

were only semi-quantitative, giving results that agreed only within $\pm 30\%$ of theoretical. Product yields using the ceric sulfate method were designated as $M'(NO_2^-)$, and those using Shinn's method were designated $M(NO_2^-)$. Cunningham observed substantial differences in product yields using the ceric sulfate technique as compared to Shinn's method for the decomposition of KNO_3 at $77°K$, but essentially no difference for KNO_3, irradiated at $300°K$. The ratio $M'(NO_2^-)/M(NO_2^-)$ being 2.0 and 1.2 respectively at the two temperatures. Sodium nitrate gave substantially different results from that of KNO_3; the ratio of $M'(NO_2^-)/M(NO_2^-)$ at $77°K$ was 3.2 and at $300°K$, it was 2.0. Annealing $NaNO_3$ and KNO_3 irradiated at $300°K$ showed about a $25\% \pm 10\%$ decrease in $M'(NO_2^-)$ yield for 2 hour annealing at $460°K$. $M(NO_2^-)$ for KNO_3 was also decreased when irradiated and annealed under these same condition but $M(NO_2^-)$ was increased for this salt when samples initially irradiated at $77°K$ were annealed at $460°K$.* The data for $NaNO_3$ were not sufficiently conclusive to warrant reporting. Initial G values were 1.70 and 0.26 for KNO_3 and $NaNO_3$ respectively. This is in contrast to Cunningham's previous values of 1.456 and 0.20.

Cunningham also did some experiments on dissolution of samples, originally irradiated in vacuo, in deaerated ceric sulfate solution. The purpose there being to prevent the reaction

$$2\,NO + O_2 \longrightarrow 2\,NO_2 \qquad (3\text{-}25)$$

from occuring with dissolved oxygen in the aqueous media. He found that $M'(NO_2^-)$ determined by this procedure did not differ by more than 15% than when $M'(NO_2^-)$ was determined by dissolution of the irradiation sample in air saturated ceric solution. The gas evolved upon dissolution of the irradiated salt in deaerated ceric sulfate solutions was collected and found to contain a small amount of nitrogen. The oxygen evolved was in the ratio $1 \pm 0.2 : 2$ to the $M'(NO_2^-)$ yield determined on the same samples. A sample of KNO_3 which had been fused while under vacuum to remove occluded gas was irradiated in vacuo and then sealed to a mass spectrometer. The

---

* It may be indicated here that Jones[82] reports that KNO_3 undergoes thermal decomposition at this temperature.

sample was heated past the transition temperatures to remove all dissolved gases and the evolved gas analyzed. The gas was found to be at least 97% oxygen. No evidences for nitric oxide or nitrogen dioxide was found in these experiments. Optical bleaching did not affect the $M'(NO_2^-)$ yields and this was taken as evidence that NO_2^{2-} or NO_3 do not contribute to $M'(NO_2^-)$. Cunningham gives an extensive discussion of his results and believes that the differences he observes using the two different methods for analysis may be attributed to the presence of NO and O_2^-. He proposes that during annealing at 460°K these species back react to form nitrite ions which accounted for the increase in (NO_2^-) he observed in the annealing studies on KNO_3. Total radical concentration as determined by EPR, using DPPH as a radical concentration standard, was determined as a function of absorbed dose. A theoretical curve was then calculated assuming the EPR signal was due to NO and that the NO concentration was given by the difference between $M'(NO_2^-)$ and $M(NO_2^-)$. The results are summarized in Figure 3-7. Although the agreement between theory and experiment is good, it is difficult to account for the fact that Cunningham observed no NO_2 or NO when the gas is analyzed mass spectrographically.

The effect of millimolar concentration of silver ion on the radiation induced decomposition of KNO_3 was studied by Cunningham.[73] The doping with silver ion was accomplished by co-crystallizing varying amounts of $AgNO_3$ with KNO_3. Samples were irradiated with gamma rays at room temperatures and analysis for NO_2^- was by the method of Shinn. Consistent with his previous studies Cunningham found that the NO_2^- yields are linear with absorbed dose and finds that initial slopes of the NO_2^- yield versus dose curves (up to an absorbed dose of about 6×10^{21} eV/g) relative to pure KNO_3 are 0.93, 0.86, 0.83 and 0.76 for salts prepared by co-crystallization of KNO_3 with 0.04, 0.1, 0.4 and 1.0 mole-% Ag^+ respectively. Beyond this absorbed dose Cunningham finds that the slopes of the NO_2^- yield versus dose curves to be the same for pure KNO_3, $KNO_3 + 0.4\%$ Ag^+ and $KNO_3 + 1\%$ Ag^+. The effect of Ag^+ on the initial yield is attributed to the scavenging of low energy electrons by Ag^+ preventing dissociative electron capture by nitrate ions. The absence of an effect of Ag^+ beyond an absorbed dose of about 6×10^{21} eV/g is explained by the equal efficiency of molecular oxygen trapped in the nitrate lattice and Ag^+ to scavenge these electrons.

Boldyrev *et al.*[74a] investigated the influence of Tl^+, Sr^{2+} and Pb^{2+} on the gamma ray induced decomposition of KNO_3. These authors observed that the decomposition yield was linear with absorbed dose with a change in slope an apparent "break", similar to that reported by Cunningham and Heal(Figure 3-1),in the curve occurring at an absorbed dose of 2×10^{21} eV/g. The effect of the addition of the ions was to increase $G(NO_2^-)$ prior to the

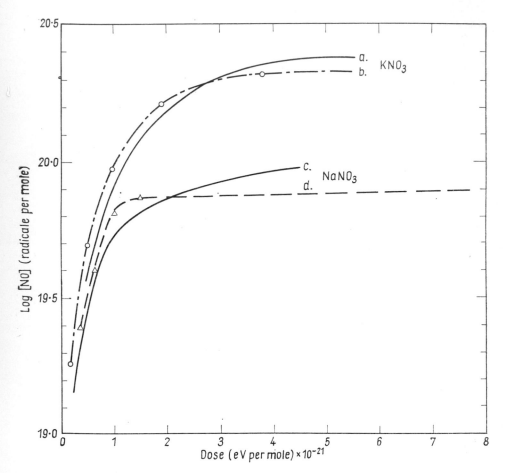

FIGURE 3-7   Production of radicals (assumed to be NO) in KNO_3 and $NaNO_3$ as a function of absorbed dose. Broken lines as estimated from e.p.r. methods, solid lines as determined by chemical methods

break and decrease $G(NO_2^-)$ (all relative to pure KNO_3) after the break. The actual reported data, however, show that this is the case for Tl^+ doping beyond 0.5 mole-% additive; for the Sr^{2+} and Pb^{2+} doped salts the effect appears appreciable only beyond 2.5 mole-% additive. The authors in agreement with Cunningham postulate the occurence of species such as NO, NO_2, NO^-, NO_2^{2-}, $N_2O_2^{2-}$ with the role of the additive acting as an electron donor or acceptor. Annealing of irradiated doped crystals indicate an accelerating effect of the additive.

Another study of LET effects in the radiolysis of sodium and potassium nitrates was reported using the $^{14}N(n, p)$ ^{14}C whose cross section for thermal neutron capture is appreciably smaller than for the nitrogen-15 containing salts. Comparison of the yields of pile irradiated nitrates containing nitrogen-15 or nitrogen-14 should, therefore, give a measure of the amount of decomposition due to pile irradiation other than neutrons since it had been previously established that no isotope effect occurs in the gamma radiolysis of the nitrates containing the ^{15}N isotope. It was also assumed in this study that more than 90% of the energy deposited as a result of the nuclear reaction would be associated with the proton liberated in the reaction. The total energy of the proton and the ^{14}C was 0.63 MeV. The LET for this reaction was estimated to be about 13 eV/Å. G values calculated on these assumptions (for the nuclear reaction only) were 0.33 ± .16 for $NaNO_3$ and 3.9 ± .6 for KNO_3. This may be compared with G values of 1.3 and 2.2 respectively for $NaNO_3$ and KNO_3 using 2 kilovolt deuterons as the radiation source. As can be seen the value obtained for $NaNO_3$ in the studies using the nuclear reaction is appreciably different from that for the alpha ray experiments[68] (LET 34 eV/Å) and 2 keV D^+ ions (LET 3 eV/Å). The ratio of the energy transferred as electronic (Ee) to that by momentum transfer (Em) was estimated and found to be 14 for $^{14}N(n, p)$ ^{14}C, 99 for 3.4 MeV alphas and 0.2 for 2 keV D^+ ions. The author concluded from Hall and Waltons[62] experiments on fission recoils that Em was less effective than Ee in causing KNO_3 decomposition and therefore since Ee predominated in the pile experiments one would have to postulate that energy transferred by electronic processes were at least as effective in producing thermal spikes as that by momentum transfer. It was felt that such a postulate was not consistent with existing thermal spike models and therefore the high G value found for KNO_3 is best explained by radical or excited specie competition in the recoil track, i.e. a model similar to that advanced in aqueous systems to explain LET effects.[75] The relatively large G value observed in KNO_3 decomposition with fission fragments and $^{14}N(n, p)$ ^{14}C events (relative to other LET values) was suggested as being due to a charge-displacement process. The charge displacement occurs via reversible electron capture-loss between the medium and the low velocity ions.

The results reported here are for relatively high doses. The discrepancy in these results with those of references 68 and 71 for the sodium salt is very appreciable. In the study under discussion samples of NaN_3 were also irradiated and the G value reported for this salt (0.7 ± 0.4) is appreciably less than that reported by Oblivantsev[154] (see Chapter V) for irradiation of NaN_3 using 4.7 MeV protons $(G = 1.7)$. In view of these substantial descrepancies in both the nitrate and azide decompositions it would appear that the calculated dosimetry in these (n, p) studies on ^{14}N and ^{15}N are in error.

A more extensive study than that reported previously[70] on the effect of pressure on the decomposition of the nitrates was made by Cunningham.[76] Helium or nitrogen gas was used as the compression fluid. All irradiations were done using cobalt-60 gamma rays. Cunningham, as a result of his previous studies on the nitrates separated the NO_2^- ion yield versus absorbed dose curve into three regions: O, I and II. For KNO_3, region 0 was defined as covering the dose ranges $0-0.4 \times 10^{21}$ eV/g: Region I about $0.5-5 \times 10^{21}$ eV/g and region II covered dose range greater than 2×10^{21} eV/g. At 77°K there was no effect of pressure (250 atm.) in region 0 for KNO_3 whether one analyzed for product by Shinn's method or by the ceric sulfate method developed in a previous paper.[72] In $NaNO_3$ irradiation at 77°K there was no effect of pressure when yields were determined by Shinn's method but a substantial effect was observed when product analyses were done using the ceric sulfate method. A 15% increase in product yield as determined by ceric sulfate analysis was observed when $NaNO_3$ was irradiated under 250 atm. pressure of helium at 77°K. It was observed that samples of $NaNO_3$ that were irradiated at 77°K under pressure and then exposed to air (during irradiation) or annealed at 300°K for 30 min. did not exhibit any difference between the two methods of analysis.

In region I a reduction in yield was observed for all samples irradiated under pressure at 300°K. This was true for samples irradiated under pure helium (700 atm.) or with helium containing oxygen (700 atm. He + 5 atm. air). Irradiation at 400°K showed an effect of the presence of air in the compression fluid; all samples ($NaNO_3$, KNO_3, $RbNO_3$, $CsNO_3$) showed an additional decrease in product yield when air was present in the compression fluid. The largest decrease was observed in $NaNO_3$ and the smallest in KNO_3. No effect of pressure was observed for decomposition in region II at 300°K for KNO_3 and $RbNO_3$. Experiments were not done in the presence of air in the compression fluid at this temperature, however, since the presence of air had no effect in region I at this temperature (300°K), it is unlikely that air should effect the decomposition in region II. In this region (absorbed dose $0 \to 5 \times 10^{21}$) the results of Cunningham are in disagreement with those of reference 69 who found no effect of pressure up to an absorbed dose of about 4×10^{21} eV/g. The effect of pressure beyond this absorbed dose was explained by Chen and Johnson as being due to the decreased probability of oxygen diffusion out of the damaged lattice which would enhance the probability of back reaction of O_2 with NO_2^- and so give a lower NO_2^- yield. However, the fact remains that a descrepancy exists between the two results which is not readily explained.

In the discussion of results Cunningham regards region 0 as that region where electron attachment processes are the most important event. In this region species such as NO_3^{2-} and perhaps other radical species such as NO may exist. Using a diffusion controlled mechanism, Cunningham explains the fact that no effect of pressure on KNO_3 in this region at 77°K was observed because of the small volume of activation. The effect of pressure on $NaNO_3$ was explained under these same conditions by assuming that since a 15% increase in product yield was observed, that this corresponded to a negative volume of activation of 3 ± 2 cm³. This negative volume of activation supports a two bond rearrangement of excited nitrate to give $NO + O_2$.

Region I is characterized by radical species (primary) reaching saturation (NO, NO_3^{2-}, NO_3) and diffusion controlled growth of secondary products by reactions such as:

$$O + NO_3^- \longrightarrow NO_2^- + O_2 \qquad (3\text{-}26)$$

$$O + NO_3 \longrightarrow NO_2 + O_2 \qquad (3\text{-}27)$$

$$O_2^- + NO_3 \longrightarrow NO_3^- + O_2 \qquad (3\text{-}28)$$

$$NO + O \longrightarrow NO_2 \qquad (3\text{-}29)$$

$$NO_3 + NO \longrightarrow N_2O_4 \qquad (3\text{-}30)$$

$$NO_2^- + NO \longrightarrow N_2O_3^{2-} \qquad (3\text{-}31)$$

occur in this region and lead to stable products. All of these reactions are diffusion controlled and hence would be affected by compression. The fact that no effect of pressure in region III is observed is explained by aggregation of dissociation fragments, partial break up of the lattice and enhanced diffusion. The effect of the presence of oxygen in the pressure fluid on the decomposition in Region 0 at 77°K was explained by oxidation of nitrogen-containing fragments, and at 400°K in Region I as being due to reaction with NO_2 or NO_2^-.

The decomposition of molten lithium nitrate was studied by Boyd and associates[77] as a function of dose rate at 270°K. The source was cobalt-60 and the dose rates employed were 8.7×10^{18}, 8.9×10^{19} and 5.6×10^{20} eV/mole-min. Dosimetry was determined by ferrous sulfate oxidation using $G(Fe^{3+}) = 15.6$. Absorbed doses were determined by multiplying the dosimeter value by the ratio of the electrons per gram of the two substances. The initial yield of nitrite ion, $G(NO_2^-)$ was the same at all three dose rates, however, beyond an absorbed dose of 0.3×10^{22} eV/mole the yield approaches a steady state and three distinct curves exist: i.e. the steady state concentration of nitrite ion was a function dose rate. The initial G value was found to be $3.73 \pm .05$. The steady state concentration of nitrite ion appeared to be lowest for the lowest dose rate but highest for the intermediate dose rate (8.9×10^{19} eV/mole-min) indicating that nitrite ion attains a maximum somewhere in the vicinity of a dose rate $= 8.9 \times 10^{19}$ eV/mole-min. Steady state nitrite values at the different dose rates are shown in Table 3-9 below.

TABLE 3-9

Dose Rate	NO_2^-
8.7×10^{18} eV/mole-min.	$1.26 \pm 13\%$ m moles/mole
8.9×10^{19} eV/mole-min.	$2.28 \pm 4\%$ m moles/mole
5.6×10^{20} eV/mole-min.	$1.77 \pm 4\%$ m moles/mole

Very little decomposition of nitrite ion was observed. A mixture of $LiNO_2$–$LiNO_3$ containing 0.924 mole-% nitrite was irradiated to a total absorbed dose of 3.13×19^{22} eV/mole without any perceptible change in nitrite ion concentration.

A kinetic mechanism was developed following a suggestion by Kostin:[78]

$$NO_3^- \xrightarrow{} NO_2^- + \tfrac{1}{2} O_2 \qquad h_1\phi \qquad\qquad (3\text{-}32)$$

$$\tfrac{1}{2} O_2 + NO_2^- \longrightarrow NO_3^- \qquad K_r\phi \qquad\qquad (3\text{-}33)$$

$$d(NO_2^-)/dt = h_1\phi(NO_3^-) - K_r\phi(NO_2^-) \qquad\qquad (3\text{-}34)$$

which on integration gives

$$(NO_2^-) = \frac{h_1 \phi}{X} (1 - e^{-xt}) \qquad (3\text{-}34)$$

where $X = h_1 \phi + K_r \phi$

and K_r = composite radiation and thermal radiation constant (min^{-1})

h_1 = radiolytic rate constant.

Equation (3-34) was shown to fit the observed data. The fitting procedure gave a single value of h_1 and three values of the composite constant K_r. The value found for h_1 was 6.19×10^{-26} mole/eV $\pm$ 1.3% and those for K_r were 4.47×10^{-4}/min (dose rate $= 8.7 \times 10^{18}$ eV/mole-min), 2.44×10^{-3}/min (dose rate $= 8.9 \times 10^{19}$ eV/mole-min) and 1.99×10^{-2}/min. (dose rate $= 5.6 \times 10^{20}$ eV/mole-min.). The authors express the opinion that the fact that $K_r = f(\phi)^n$ with $n > 1$ implies that at least two species, whose concentration are at least in part radiation produced, must be involved in the back reaction. A post irradiation thermal reaction to produce nitrate was observed which amounted to a 10% reduction in nitrite ion concentration in about 50 minutes. It may be added here than Cho and Johnson[79] found that the activation energy for the thermally induced back reaction (equation 3-33) was only 7.8 $\pm$ 2 kcal/g. mole whereas that for the forward reaction was 26.2 $\pm$ 1.4 kcal/g. mole. Hence it would appear that steady state oxygen concentrations, aside from diffusion considerations, would in all cases be very low.

In a subsequent paper, Boyd *et al.*[80] studied the decomposition of crystalline and molten lithium nitrate using recoil particles from the $^6Li(n, d)\,^3H$ reaction. The recoil particles from this reaction share a kinetic energy equal to 4.79 MeV, and have an initial LET of approximately 30 eV/Å. Samples of pure $^7LiNO_3$ were similarly irradiated and the results obtained with this salt gave a good measure of the extent of decomposition due to fast neutrons and pile gamma-rays. Variation in dose rate for this system could be achieved by varying reactor power or the isotopic fraction of the lithium-6.

Samples of $LiNO_3$ containing 0.513 and 1.015 atom-% of ^6Li were irradiated at 40° and 270°C. At 40°C the nitrite yield was linear with absorbed dose up to an absorbed dose of about 6×10^{23} eV/mole. $G(NO_2^-)$ was found to be 0.22 and $G(O_2)$ was 0.11. No effect of varying the dose rate by a factor of about 4 was observed. The G value found by Hochanadel using 3.4 MeV alpha particles was 0.7 even though the LET was about the same. The difference between the two values was seen to indicate either, (1) a dose rate effect (the dose rate in Hochanadel's experiments were 60 times that reported here) or, (2) a small amount of melting occurred during the alpha-ray experiments. In the molten state (270°C) irradiations were only done on the salt containing 1.015 atom percent of ^6Li. The initial G value was found to be 5.5 $\pm$ 0.5 and the extrapolated steady state concentration of NO_2^- was found to be 33 mMoles per mol $LiNO_3$. The data were found to fit equation (3-34). More than 95% of the gas liberated during decomposition was found to be oxygen; small amounts of N_2, CO, NO, N_2O, and CO_2 were detected. Measurements of hydrogen ion concentration of solutions of irradiated molten salt indicated that small amounts of Li_2O were also found (about 0.1 mMole/mole $LiNO_3$). The G value observed for the salt in these experiments was considerably higher than those observed in the gamma-ray experiments. The authors suggest that perhaps this may be due to competition between reactions:

$$NO_2^- + O \longrightarrow NO_3^- \qquad (3\text{-}35)$$

$$O + O \longrightarrow O_2. \qquad (3\text{-}36)$$

The smaller O atoms it is believed will diffuse out of the track faster than the nitrite ions hence reaction (3-36) would be favored. Another possibility which may be included here are complex reactions of the oxide with the nitrate or of oxides of nitrogen (NO, NO_2, etc.) with nitrite[81] which may be produced in the track.

Ladov and Johnson[36] have examined the oxygen isotope effect in the decomposition of KNO_3. Samples of $KN^{16}O_3$ and $KN^{18}O_3$ were irradiated using a cylindrical cobalt-60 source at different temperatures. Dosimetry was done using ferrous sulfate oxidation with $G = 15.45$. Nitrite was determined by Shinn's method using a molar extinction at $\lambda = 536$ mμ of 52,000. The $KN^{18}O_3$ was used as purchased without further purification since analysis by the National Bureau of Standards showed that the principal impurities were Na(0.01%) and Ca(0.006%). The results for irradiation at room temperature are summarized in Figure 3-8 and Table 3-10. As is apparent from Figure 3-8, the initial slope for the isotopically enriched salt is distinctly different from that of the normal salt, however, beyond an absorbed dose of about 4×10^{21} eV/g the slopes are roughly parallel. The isotope effect increased up to an absorbed dose of about 0.5×10^{21} eV/g after which it decreased and reached a constant value of $9 \pm 3\%$. Cunningham[63] had reported that yields of NO_2^- were linear with absorbed dose and the results could be plotted as two straight lines intersecting at an

TABLE 3-10   Decomposition of normal and isotopically enriched (70.5%) KNO_3[a]

Dose, eV/g $\times 10^{-21}$	Molecules of Product/g $\times 10^{-19}$				Ratio $G(N^{16}O_2^-)/G(N^{18}O_2^-)$
	$N^{16}O_2^-$	$N^{18}O_2^-$	$G(N^{16}O_2^-)$	$G(N^{18}O_2^-)$	
0.101	0.141	0.136	1.40	1.35	1.04 ± 0.02
0.104	0.145	0.137	1.39	1.32	1.06 ± 0.02
0.207	0.289	0.267	1.40	1.29	1.08 ± 0.03
0.351	0.479	0.425	1.36	1.21	1.12 ± 0.03
0.445	0.608	0.509	1.37	1.14	1.19 ± 0.03
0.451[b]	0.619	0.523	1.37	1.16	1.18 ± 0.03
0.807	1.08	0.925	1.34	1.15	1.16 ± 0.03
0.807	1.07	0.936	1.33	1.16	1.15 ± 0.03
1.27	1.64	1.40	1.29	1.10	1.17 ± 0.03
1.33	1.69	1.47	1.27	1.11	1.14 ± 0.03
2.38	3.02	2.62	1.27	1.10	1.15 ± 0.03
4.07	4.87	4.33	1.20	1.06	1.12 ± 0.03
10.7	10.3	9.33	0.963	0.872	1.10 ± 0.03
12.6	11.8	10.8	0.937	0.857	1.09 ± 0.03
18.9	16.6	15.2	0.878	0.804	1.09 ± 0.03

[a] Irradiated at room temperature (20°C)
[b] Irradiated in vacuo.

absorbed dose of about 7×10^{21} eV/g, which gave rise to the apparent "break" referred to previously. He observed an isotope effect of about $12 \pm 2\%$ up to the break and an isotope effect of about $2 \pm 7\%$ after the break. In agreement with previous results[42] Ladov and Johnson find that the decomposition is not linear with absorbed dose indicating that back reaction was occuring in this salt. Substantial proof for the occurrance of back reaction was obtained from examination of the Raman spectra of $KN^{18}O_3$ irradiated in the presence of air. Figure 3-9 shows the Raman spectra of the isotopically enriched salt before and after irradiation to an absorbed dose of about 8×10^{21} eV/g. The four peaks shown in Figure 3-9 are due to the symmetric stretching frequencies of NO_3^- ion containing 3 oxygen-18 atoms, 2 oxygen-18 atoms, plus one oxygen-16 atom, one oxygen-18 atom plus two oxygen-16 atoms and 3 oxygen-16 atoms. The growth of the peak due to $N^{16}O_3^-$ is apparent and unequivically established the occurence of a radiation induced reaction of NO_2^- with molecular oxygen. The effect of impurities on the decomposition of KNO_3 was investigated to eliminate

5*

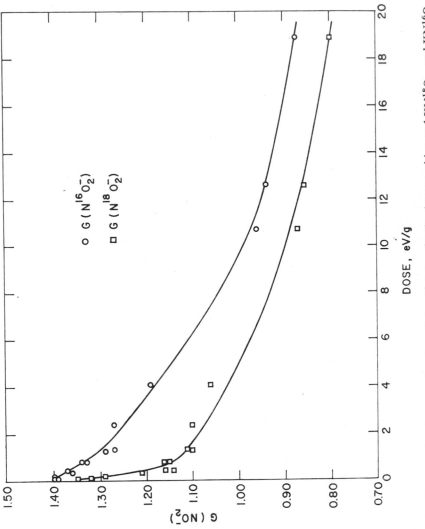

FIGURE 3-8   *G* values as a function of absorbed dose for the decomposition and $KN^{18}O_3$, and $KN^{16}O_3$

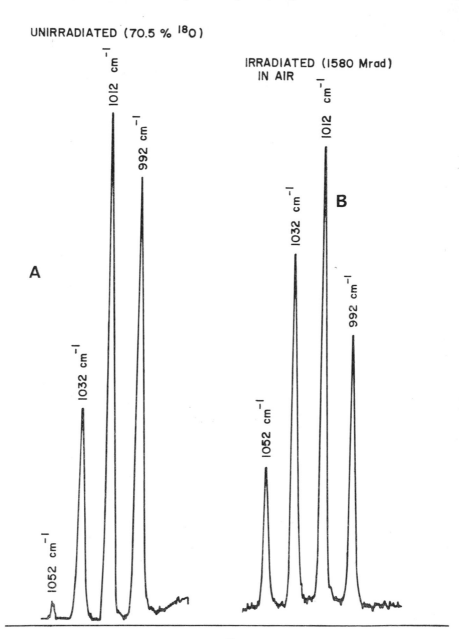

FIGURE 3-9   Raman spectra of $KN^{18}O_3$ before and after irradiation in air

the possibility of the isotope effect being due to cation or anion impurities incorporated in the isotopically enriched salt during preparation.* Samples of normal KNO_3 crystals were recrystallized from water containing excess acid or base and then dried in a vacuum oven at 100°C. No change in $G(NO_2^-)$ was observed. Since the isotopically enriched salt contained calcium ion as an impurity, samples of KNO_3 containing 0.04% Ca^{2+} and 1% Ca^{2+} were prepared. No effect of addition of 0.04% Ca^{2+} ion was found, however, a 10% decrease in $G(NO_2^-)$ was observed for the sample containing 1% Ca^{2+}. This is about the magnitude observed by Cunningham for the addition of Ag^+ (16%). However, if $G(-KNO_3)$ was calculated on the basis of the electron fraction of KNO_3 in the mixture, the effect of Ca^{2+} was a few percent.

Cunningham had observed that KNO_3 irradiated at 77°K and dissolved without warming in acid ceric sulfate solution produced a reduction in ceric ion in excess of that due to the presence of nitrite ion. He attributed this additional reduction of ceric ion to the presence of NO and possibly O_2^-. This effect was explained by Ladov and Johnson as being due to the well known reactions of electron excess and deficient centers with the aqueous media to produce H atoms and OH radicals respectively. The H atoms reduced ceric to cerous and the OH radicals combined (in the absence of cerous) to form H_2O_2 which reduces ceric.

Ladov and Johnson explained the abnormally high isotope effect in the initial stages of the decomposition as follows: they proposed that decomposition occurred via two reaction paths; reaction path I was decomposition by excitation and ionization (essentially equations 3-8, 3-9, 3-10) and path II was decomposition via energy transfer at certain preferred lattice sites. The high initial isotope effect was attributed to the sum of these two decomposition modes. After an absorbed dose of about 0.5×10^{21} eV/g decomposition via energy transfer ceases due to the loss of preferred lattice sites and/or exciton scavenging by molecular oxygen and the isotope effect slowly falls to the steady state value of 9 ± 3%. The isotope effect at this point being due to decomposition via reaction path I.

Contrary to the results reported by Cunningham no effect of temperature was found on the isotope effect over the range 77–373°K. These results are summarized in Table 3-11.

---

* The salt is prepared using extremely pure materials. The method of preparation is exchange of HNO_3 with $H_2^{18}O$, neutralization of the HNO_3 with KOH and subsequent recrystallization from enriched water in the presence of HNO_3.

TABLE 3-11   Isotope effect as a function of temp.

Temp.	Absorbed Dose	$G(N^{16}O_2^-)/G(N^{18}O_2^-)$
77°K	0.224×10^{21} eV/g	1.12
77°K	0.268×10^{21} eV/g	1.14
293°K	0.207×10^{21} eV/g	1.12
373°K	0.252×10^{21} eV/g	1.10

The isotope effect originating with the energy transfer process was explained as follows: oxygen produced by reaction path I undergoes a random diffusion. When it arrives at one of the preferred locations (lattice sites) reaction occurs and that particular site can no longer function as a decomposition site. The "preferred" sites were thought to be equivalent to the trapping sites found in the alkali halides because the number of such sites in KNO_3 as estimated from the initial isotope effect, was found to be about 10^{17}–10^{19} cm^{-3} which was the same order of magnitude obtained for electron excess centers in the alkali halides. Possible reactions at these "preferred" sites were:

$$[NO_3^-] + O_2 \xrightarrow{h\nu} NO_3 + O + O^-$$

$$[NO_3^- + O] \longrightarrow NO_2^- + O_2$$

$$O + NO_3^- \longrightarrow NO_2^- + O_2$$

or $\qquad [NO_3^- + O_2] \xrightarrow{h\nu} NO_2^- + O_2 + O.$

It was postulated by these authors that all decomposable oxyanions (bromates, chlorates, perchlorates) should show an isotope effect. In tightly bound lattices such as $NaNO_3$ or at very low temperatures where diffusion would be restricted the isotope effect would be restricted to that occurring via reaction path I.

Jones[82] has studied the isotope effect in radiation induced decomposition of $KN^{18}O_3$. He observed a 13% isotope effect up to an absorbed dose of 4×10^{21} eV/g which was verified by analysis of the oxygen gas produced during the decomposition. He also studied the decomposition of KNO_3 doped with varying amounts of KNO_2 and established that oxygen yields fell off sharply with increasing KNO_2 concentration. He attributed this fall off in oxygen as due to the radiation induced back reaction of oxygen with the added NO_2^-. The decomposition of KNO_3 above 150°C was also studied.

The yield of NO_2^- was independent of temperature if corrections were made for thermal decomposition of KNO_3.

The effect on the decomposition of sodium nitrate of doping with varying amounts of calcium nitrate was studied by Kaucic and Maddock.[82a] This particular system was chosen over others because of the thermal stability and homogenity of the solid solutions so produced, when varying amounts of $Ca(NO_3)_2$ were added to molten $NaNO_3$. The decomposition products were analyzed for by the method of Shinn and by reaction with acid ceric sulfate solution. This latter method was used by Cunningham with the express purpose of detecting reducing species other than NO_2^- (see reference 72). The yield of NO_2^- as determined by the method of Shinn was designated as $M(NO_2^-)$ whereas that determined by the ceric sulfate method was designated as $M'(NO_2^-)$.

The ratio of $M'(NO_2^-)/M(NO_2^-)$ as a function of absorbed dose for pure $NaNO_3$ and $NaNO_3$ containing varying amounts of $Ca(NO_3)_2$ up to 0.5 mole-% had an initial value of about 2.5 but approached a value of unity beyond an absorbed dose of about 1×10^{21} eV/g (Figure 3-10). However, doping increased the initial $G(NO_2^-)$ values (as determined either by Shinn's method or ceric sulfate). $G(NO_2^-)$ increased with increasing Ca^{2+} concentration up to 0.1–0.2 mole-% Ca^{++} concentration. At this point $G(NO_2^-)$ reached a maximum and became constant. $G(NO_2^-)$ at the maximum was increased by about 50% by the addition of Ca^{2+}.

The authors explained the effect of doping on $G(NO_2^-)$ as being due to characteristics of the cation vacancies caused by the addition of Ca^{2+} and the shallower electron traps presented by NO_3 radicals tied to such a vacancy.

Other explanations for the effect of Ca^{2+} on the decomposition are of course valid. In KNO_3, Ladov and Johnson found a decrease in $G(NO_2^-)$ as a result of added Ca^{2+} and were able to account for this decrease on the bases of the amount of energy absorbed by the $Ca(NO_3)_2$ since the G value for $Ca(NO_3)_2$ decomposition is about one-third of that for KNO_3. This interpretation seemed reasonable since the decrease in $G(NO_2^-)$ appeared to be proportional to the Ca^{2+} concentration. A similar type calculation in the $NaNO_3$ system would not account for the substantial increase (about 50%) in the $G(NO_2^-)$ observed. However, Kaucic and Maddock do observe a substantial increase in $G(NO_2^-)$ for a purely mechanical mixture of $Ca(NO_3)_2$ and $NaNO_3$.

Another way to look at this system is to consider $Ca(NO_3)_2$ as an impurity in the $NaNO_3$ lattice. Application then of the energy transfer principle could

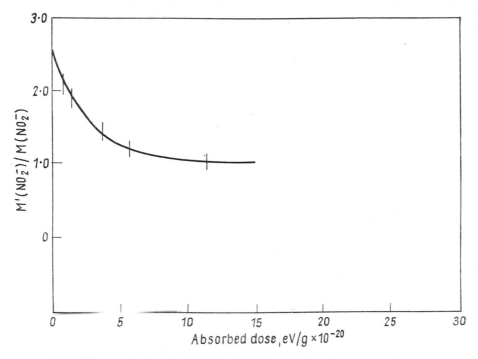

FIGURE 3-10   Ratio of $M'(NO_2^-)/M(NO_2^-)$ as a function of absorbed dose

readily account for the increase in G upon addition of Ca^{2+}. The calcium ion is large and will certainly distort the $NaNO_3$ lattices and conceivably may act as a trapping site. This does seem reasonable considering the fact that the effect on $G(NO_2^-)$ by Ca^{2+} reaches a maximum at relatively low concentrations of Ca^{2+}.

Most interesting, however, in these studies is the ratio $M'(NO_2^-)/M(NO_2^-)$ which appears to level off at unity (both for the pure salt and the doped salt) at an absorbed dose of about 0.5×10^{21} eV/g. This, it may be recalled, is the magic number discussed in Chapter 2 under the section "energy transfer". The ratio of $M'(NO_2^-)/M(NO_2^-)$ for the room temperature radiolysis of KNO_3 is approximately unity hence the additional "reducing species" found in $NaNO_3$ are unique to this salt and are not dependant upon the presence of an impurity atom. The additional reducing species detected by the ceric sulfate method are not identified and may simply be trapped electrons and holes. Cunningham observed a 50% decrease in $M'(NO_2^-)$

when $NaNO_3$, initially irradiated at 77°K, was annealed at 460°K for two hours and a 25% decrease in $M'(NO^-)$ for a two hour anneal at 460°K for samples irradiated at 300°K.

3.2   Optical absorption bands

Almost all of the nitrates become colored when exposed to ionizing radiation or UV. light (5320 Å) however, there has been relatively few studies on the properties of the absorption bands so produced. Pringsheim[83] studied optical absorption band formation in $NaNO_3$ irradiated with cobalt-60 gamma rays and X-rays. He observed the presence of two bands, one due to the presence of nitrite ions with an absorption peak at 345 mμ and another attributed to a color center band at 335 mμ. The color center band was readily bleached at −190°C, and was decomposed by irradiation at room temperature with X- or gamma-rays. The proximity of the color band peak to that of the nitrite ion suggested that the color center band was due to NO_2^- ions strongly perturbed by the capture of an electron in their vicinity.

Cunningham[84] studied the radiation induced optical absorption in KNO_3 crystals irradiated with cobalt-60 gamma-rays and X-rays at 77°K and 4.4°K. He identified seven radiation induced optical absorption bands and related them to EPR spectra of similarly irradiated KNO_3. The position of the maxima of each of the observed bands was 510 mμ, 355 mμ, 290 mμ, 270 mμ, 241 mμ, and 410 mμ. The bands at 510 and 270 mμ were both bleached when exposed to 510–680 mμ light and they were identified with the species NO_2^{2-}. The band at 410 mμ appeared only after exposure to relatively large doses ($\sim 10^{20}$ eV/g) and was assigned to the NO_2 radical. The band at 246 mμ was assigned to the species O_2^- because of the similarity of the properties of this band to that of the EPR signal with respect to temperature i.e. this band did not appear at 4.2°K but did appear when the crystal was heated to 77°K, an effect observed in the EPR studies. The band at 325 mμ was attributed to the species NO_3^{2-}. The bands at 355 and 290 were not assigned to any particular species but were thought to be related in some complex fashion to the species identified as NO_2, NO_2^{2-}, or NO_3^{2-}.

Some relevant studies on color centers in alkali halides doped with nitrites and nitrates have been published.[85] It has been observed that in irradiated KCl, KBr, NaBr and NaCl doped with $NaNO_2$, F center bands are completely suppressed at room temperature at high NO_2^- concentrations.

A C band is produced which has a continuous absorption band with a wavelength that is nearly coincident with the NO_2^- band. The C band is believed to be an electron excess band.

3.3 Paramagnetic Species

There have been many studies on paramagnetic species in irradiated nitrates. Two recent reviews are available which are concerned with paramagnetic species in all inorganic solids.[86,87] Ard[88] studied the paramagnetic resonance of color centers in $NaNO_3$ irradiated at room temperatures and observed the presence of two sets of lines. One set consisted of three equal-intensity, equally spaced lines with a total separation of 135 ± 5 gauss. These lines were interpreted as hyperfine structure due to coupling with the ^{14}N nucleus which he attributed to either the NO_2 or NO_3 radical. The other set of lines changed with crystal orientation in the magnetic field. When the longer crystal diagonal is parallel to the external field three lines are observed at $g = 2.013, 2.020$ and 2.026. These lines have a width of about 3 gauss. When the shorter diagonal is parallel to the external field one line is observed at a g value of 2.00. Ard indicated that the 3 gauss width for the lines associated with this center made it unlikely that they could be attributed to F centers.

Bleany, Hayes and Llewellyn[89] studied the paramagnetic resonance spectrum of lanthanum magnesium nitrate, $La_2Mg_3(NO_3)_{12} \, 2 \, 4H_2O$. The paramagnetic centers were formed by radiation from americium-241 (alpha particles) and from promethium-147 which had been separately co-crystallyzed with the lanthanium magnesium nitrate. When the external magnetic field is parallel to the crystallographic c-axis the spectrum consists of a three line hyperfine structure which is due to the nitrogen nucleus. They eliminated N, NO, NO_3 and NO_2^{-2} as being responsible for the observed centers and indicated that although NO_2^{-2} could give a hyperfine structure of the appropriate magnitude it had the wrong anisotrophy. Hence they attribute the magnetic centers observed to NO_2. These authors also examined the spectrum of $NaNO_3$, irradiated with gamma-rays and found a complicated spectrum but did not observe a nitrogen hyperfine structure. A sample of $(La, \, 10\% \, Ce)_2 \, Mg_3(NO_3)_{12} \, 24 \, H_2O$ irradiated with gamma-rays gave the same spectrum as that for salt containing no added cerium.

Zeldes and Livingston[90] irradiated single crystals of $NaNO_2$ at $77°K$ and observed an anisotropic three line hyperfine spectrum associated with a single paramagnetic species which they believed to be due to NO_2.

Jacard[91] has observed the presence of paramagnetic nitrogen oxides in irradiated alkali halides doped with nitrate or nitrite (0.01–1 wt. %). He observed six triplets and one singlet in KCl doped with NO_3^-. One of the centers was attributed to NO since its concentration was increased by heating the pure crystal in NO_2 gas or in pure NO gas, and another believed to be interstial NO. Another center was attributed to nitrate ion in a negative ion vacancy trapping an electron during irradiation. The intensity of this specie was greatly reduced by hole capture and hence was assigned to NO_3^{2-}. A center, produced only in crystals doped with NO_2^- was observed and tentatively assigned to NO_2^{2-}. The singlet which was observed was attributed to oxygen formed during the decomposition of nitrate. The spectrum did not appear in crystals heated in an atmosphere of NO or NO_2. The intensity of the spectrum associated with this specie increased with increased absorbed dose and was attributed to O_2^+ localized at a neighboring chloride ion.

Cunningham[92] studied the paramagnetic centers in irradiated $K^{14}NO_3$. The species identified by Jacard as NO_3^{2-} was identified by Cunningham as NO_2^{2-}. The specie identified by Cunningham as NO_2^{2-} was only observed for short irradiation times. Longer irradiation times produced a triplet at a greater spacing than the NO_2^{2-} and was therefore attributed to NO_2 radicals. Another triplet was assigned to NO since his values for the g and A tensor were very similar to the values found by Jacard and attributed by him to NO. A triplet only observed by irradiation at 4°K was assigned to NO_3^{2-}. A signal at $g = 2.006$ was observed at 4°K which appeared to have no measurable hyperfine splitting and with cylindrical symmetry of its g-factor about the C-axis was assigned to the NO_3 radical. A single isotropic line which appeared on warming crystals irradiated at 4°K to 77°K was attributed to O_2^-.

Zeldes[93] examined the paramagnetic species in irradiated KNO_3. The parameters attributed to NO_2^{2-} by Cunningham were assigned to NO_3^{2-} by Zeldes. The parameters obtained by Zeldes for this specie were almost identical with those attributed to it by Jacard. Other species which were considered and rejected were N, NO and NO_3. The only other specie found was NO_2. Some of the principal values of g and hyperfine values A were significantly different from those found in $NaNO_2$ but the striking similarity of the traces and the similarity of their average values led him to assign this spectrum to NO_2. The difference in the individual parameters, is believed due to the motion of NO_2 in the KNO_3 lattice. Zeldes however, observed that the spectrum for NO_2 did not appear unless NO_2^- was present in the lattice. If pure KNO_3 is irradiated at $77°K$ the spectrum of NO_2 does not appear, however, if pure KNO_3 doped with KNO_2 is irradiated at $77°K$ then the signals for NO_2 appear. When pure KNO_3 irradiated at $77°K$, is warmed to room temperature, and then reirradiated at $77°K$ the NO_2 signal is observed. Thus it appears that NO_2 is only observed when NO_2^- is present in the lattice and furthermore that NO_2^- is not formed directly at $77°K$. Zeldes further suggests that NO_2 is produced in anion vacancies by ionization of NO_2^- in NO_3^- ion vacancies.

Zeldes and Livingston[94] examined the paramagnetic spectra of irradiated KNO_3 containing about $1\% KNO_2$, 95% enriched in nitrogen-15. The NO_2 spectra was observed and the ratio of the intensity of the lines for $^{15}NO_2$ to those for $^{14}NO_2$ is just that expected from the 95% ^{15}N content in the added KNO_2. All the NO_2 observed therefore, comes from the added NO_2^-. The authors conclude therefore, that no NO_2^- is produced on irradiation of KNO_3 at $77°K$ but a precursor is formed which then gives NO_2^- on warming.

Golding and Henchman[95] irradiated lead nitrate at room temperature and observed the presence of three paramagnetic species. Although irradiations were done at room temperature the spectra were examined at $77°K$. One specie was identified as NO_3. The basis for the assignment was the absence of marked nitrogen hyperfine splitting and the similarity of the g tensors to those reported by Zeldes.* The other specie identified was NO_2. The g tensors were identical to those reported by Zeldes and Livingston[94] and the hyperfine splitting constants were such as to indicate that the molecule is rotating about its axis.

Cunningham, McMillan, Smaller and Yasaitis[96] irradiated KNO_3 enriched to 99.6% in nitrogen-15 and studied the paramagnetic spectra at 4.2 and $77°K$. They observed four triplet signals and one singlet signal. Three of the triplets were tentatively assigned to NO_2^{2-}, NO and O^- and one triplet signal was unidentified. The singlet was assigned to O_2^-. The data reported in this paper is essentially the same as that in reference 92.

The formation of the NO_2 radical at room temperature has also been observed in $Sr(NO_3)_2$.[97] Zdansky and Sroubek observed the presence of four paramagnetic centers in $Sr(NO_3)_2$ irradiated at room temperature. One of these species they identified as NO_3^{2-}. This center appeared after only a short irradiation time and the intensity of the signal was proportional to absorbed

---

* Private communication.

dose, however, the signal saturated at a concentration of centers equal to about 10^{16} cm^{-3}. After prolonged irradiation three other paramagnetic centers were observed all of which had a much lower intensity than that attributed to NO_3^{2-}. By bleaching and thermal annealing processes the signals were isolated. One of the signals was attributed to NO_2. The intensity of the spectrum responsible for this specie was reduced by UV bleaching but appeared to remain unchanged by annealing at 200°C; it was increased substantially when $Sr(NO_3)_2$ doped with $NaNO_2$ was irradiated. The other two centers could not be resolved, however, the species NO_3 and O_3^- were suggested.

A subsequent paper by these same authors[98] on the paramagnetic species in irradiated strontium and lead nitrates reports on the presence of NO_3^{2-} but does not discuss any other paramagnetic species. The values obtained for the parameters for the observed spectrum were very similar to those reported by Jacard and Zeldes for NO_3^{2-}.

Boesman *et al.*[99] have studied the paramagnetic species in irradiated single crystals of the alkali halides doped with $AgNO_3$, KNO_3 and $NaNO_3$. All spectra were examined at room temperature. In these systems they have identified NO_2 and NO_3, however, the hyperfine structure constants do not agree with those of Zeldes or Jacard.

Addé and Petit[100] observed two paramagnetic spectra when $NaNO_3$ was irradiated with 1.5 MeV electrons to an absorbed dose of 10–250 Mrad. The precision of the measurements did not permit resolution of the anisotrophy of one of the spectra. The other spectra because of the similarity of the g factor and hyperfine structure constants with those of Livingston and Zeldes was attributed to NO_2.

Geisi and Kazumata[101] observed the spectrum of gamma irradiated $NaNO_3$ at room temperature and 77°K. The room temperature spectrum consisted of four sets of lines. One spectrum was a triplet while the other three could not be resolved. A sample irradiated at 77°K and examined at this temperature produced an intense singlet. The g value of the singlet did not change during rotation of the magnetic field around the 111 axis. They attributed this spectrum to the neutral NO_3 lying in the 111 plane and probably at the original NO_3^- site. The triplet observed a room temperature was attributed to the NO_2 radical. None of the other spectra were resolved.

A rather thorough study of the ESR spectrum of irradiated silver nitrate by Mosely and Moulton[102] indicated the presence of Ag^{2+}, $AgNO_3^-$ and

NO_2. Another paramagnetic species (a singlet) was observed, however, identification was not possible. Studies were made at room temperature and at 77°K. The spectra of Ag^{2+} and $AgNO_3^-$ were only observed at 77°K; warming the irradiated crystal to 135°K causes an irreversible loss of both spectra. The spectrum attributed to NO_2 was only observed if NO_2^- was present in the lattice. The intensity of the NO_2 spectrum was found to decrease on continued irradiation and was not detected above 240°K. As with $NaNO_3$, when irradiated pure $AgNO_3$ is warmed to room temperature, cooled to 77°K and reirradiated the spectra of NO_2 appears indicating that NO_2^- is not formed directly, but does so from some precursor. Attempts were made to determine the exact temperature at which nitrite appears in the lattice using the NO_2 spectrum as an indicator. Pure $AgNO_3$ irradiated at 77°K was warmed to 135°K. The spectra due to Ag^{2+} and $AgNO_3^-$ disappeared, however, on cooling to 77°K and reirradiating the spectra of NO_2 does not appear, thus indicating that $AgNO_3^-$ is not the precursor for NO_2^-. Annealing above 200°K did show, after subsequent irradiation, the spectrum of NO_2. However, it was also observed that in a sample containing NO_2, warming causes the spectrum of NO_2 to increase while that of Ag^{2+} (which disappears at a slightly lower temperature than $AgNO_3^-$) to decrease. This suggests that NO_2 might arise by capture of an electron by Ag^{2+}.

A study on electron hole trapping in calcite and $NaNO_3$ crystals irradiated at 77°K supported previous paramagnetic studies on the latter salt.[103] The species NO_3^{2-} and NO_2 (rotating) were observed. In addition an intense singlet was observed which was attributed to NO_3.

A study of the paramagnetic spectrum of NO_2 formed during the low temperature radiation of potassium nitrate[104] revealed that the thermal motion of NO_2 in this lattice is frozen out at 4°K. Previous studies by Zeldes[93] have shown that some of the principal values of g and A for this specie departed significantly from that of NO_2 in the $NaNO_2$ lattice. To account for this it was proposed that the NO_2 in the KNO_3 lattice was in a NO_3^- vacancy and was rapidly reorienting between sites related by a mirror plane. In the studies reported here it was observed that the NO_2 lines broadened and became extinct as the temperature was reduced, however, as the temperature approached 4°K the lines reappeared at different positions; the temperature effect was reversible. At 4°K the g and hyperfine parameters are essentially the same as those for NO_2 in $NaNO_2$, but substantially different than in KNO_3 at 77°K.

3.4   Summary and critical review of nitrate data

3.4.1   Optical spectra

There are only three significant papers on the optical spectra of irradiated nitrates. The work of Pringsheim appears to indicate the presence of a specie absorbing in the vicinity of the nitrite ion with several sattelite peaks. On warming the sattelite peaks decay and the spectrum of NO_2^- appears. There is evidence for a color center which Pringsheim attributes to NO_2 perturbed by a neighboring electron.

With KNO_3 a similar situation appears to exist, however, Cunningham has interpreted the spectra as consisting of several bands which he believes are due to the presence of various species including NO_2^{2-}, NO_3^{2-}, NO_3, NO, and O_2^-. ESR studies support the existence of some of these species however, considering their concentration (as determined by ESR techniques) the extinction coefficients must be quite high.

3.4.2   Paramagnetic species

From the review of the literature on paramagnetic species it appears that the following radicals are present in irradiated nitrates:

NO_2:   There is general agreement concerning the presence of this specie in irradiated nitrates. This radical has been observed at low temperatures in $NaNO_3$, $NaNO_2$, KNO_3, $AgNO_3$, and at room temperature in $Sr(NO_3)_2$ and $Pb(NO_3)_2$. In the $NaNO_2$ lattice it is stationary while in the KNO_3, $Sr(NO_3)_2$, $Pb(NO_3)_2$ it is rotating. In $Pb(NO_3)_2$ and $Sr(NO_3)_2$ the formation of NO_2 does not require the prior presence of NO_2^- in the lattice whereas in $AgNO_3$, $NaNO_3$ and KNO_3 it does. However the signal for NO_2 is greatly enhanced in $Sr(NO_3)_2$ if NO_2^- is present in the lattice.

NO_3^{2-} With the exception of $AgNO_3$, all other nitrates examined ($NaNO_3$, KNO_3, $Sr(NO_3)_2$, $Pb(NO_3)_2$) show the presence of this specie. Although Golding and Henchman did not observe this specie it must be remembered that in their case the salt was irradiated at room temperature and examined at 77°K.

NO_2^{2-} This specie has only been observed by Cunningham in his earlier studies. However, the parameters assigned by him to this specie are those which Zeldes, Zdansky, Jacard and Cunningham (in a later paper) assign to NO_3^{2-}.

	g (gauss)		Hyperfine splitting Constants (gauss)		Reference
NO_2^{2-}	2.0060	2.001	43.2	62.5	92
NO_3^{2-}	2.0057	2.0015	31.8	63.4	93
	2.0068	2.0020	30.5	61.5	91
	2.0060	2.0019	37.3	68.8	98
	2.0067	2.0019	33.8	64.4	103

We may conclude that the evidence for the presence of NO_2^{2-} in irradiated nitrates is not sufficient to warrant its consideration as a primary specie.

NO_3 This specie has been observed in $NaNO_3$, KNO_3, and $Pb(NO_3)_2$. The specie that Boesman *et al.* assign to NO_3 is that which Zeldes[93] and Jacard,[91] assign to NO_3^{2-}. This specie may also be present in $Sr(NO_3)_2$ and $AgNO_3$ although no assignment in studies on these compounds was made.

NO   The only evidence for this specie is that reported by Cunningham[92] on irradiated KNO_3. Zeldes, Zeldes and Livingston and Holmberg and Livingston have all examined the ESR spectra of irradiated KNO_3 and find no evidence for its presence.

O_2^-, O   These species have been reported by Cunningham[92] however, as with NO there is no corroborating evidence from other workers.

3.4.3   *Mechanism of the decomposition*

We conclude from all the available evidence that NO_3^{2-}, and possibly NO_3 are primary specie in the radiolysis of the nitrates. Except for $Pb(NO_3)_2$, and $Sr(NO_3)_2$, NO_2 does not appear unless NO_2^- is present. We do not therefore consider it a primary specie. Furthermore nitrite ion does not appear to form directly but via some intermediate specie. The primary act may therefore be visualized as follows:

$$NO_3^- \rightsquigarrow \begin{bmatrix} \text{electron} \\ \text{excess} \\ \text{centers} \end{bmatrix} + \begin{bmatrix} \text{electron} \\ \text{deficient} \\ \text{centers} \end{bmatrix} \longrightarrow NO_2^- + \tfrac{1}{2}O_2 \qquad (3\text{-}37)$$

$$\longrightarrow NO_3^- \qquad (3\text{-}38)$$

It is pretty well established that the concentration of paramagnetic species saturates i.e. beyond a certain absorbed dose the concentration of these species is constant. This being true if some sort of electron excess or deficient center is the precursor for NO_2^- then it would appear that G values for irradiations done at 77°K would be less than at 300°K. This is not the case. Furthermore, Pringsheim visually observes the nitrite ion (with some satellite structure) in $NaNO_3$ at 77°K. A possible explanation then for the fact that NO_2 is only observed if NO_2^- is present in the lattice in KNO_3 is that the precursor for NO_2^- is not a paramagnetic specie or that NO_2 can only arise from NO_2^- that is formed at some particular lattice site (dislocation, vacancies etc.). We would expect if this is true that at low temperature and high absorbed dose the NO_2 signal would be observed. As the matter now stands this question will have to await further experimentation.

Although we may consider reactions (3-37) and (3-38) to be the significant primary reactions there must be others. The decomposition of $AgNO_3$ and $Pb(NO_3)_2$ are known to be more complex than simple decomposition to give nitrite ion and oxygen. In $AgNO_3$ the species $AgNO_3^-$ and Ag^{2+} have been detected, however, their role in $AgNO_3$ decomposition is not known. Other nitrates which have not been studied as yet may show similar complex behavior. Until such time, however, we consider reactions (3-37) and (3-38) or reactions very similar involved in the initial act. The observation of Cunningham and Heal that nitrite is formed at random lattice sites, (as established by X-ray diffraction and IR studies) and not at some localized areas must be accepted. In addition the argument by Ladov and Johnson that an energy transfer mechanism is operable in the initial stages of the decomposition should be considered as should the existence of a radiation induced back reaction. There does not appear to be any effect of intensity on the decomposition hence competition reactions of primary or intermediate species do not appear to be too significant. This observation, and the results of Hall and Walton would support the view of Hochanadel that LET effects may be explained on the basis of the production of high local temperatures, not on the basis of competitive reactions of primary or intermediate species. To summarize:

1)   Initial act to give excited species which can decompose to yield nitrite plus oxygen or produce NO_2 plus the oxide (or free metal plus O_2 etc.). Reaction occurring at random lattice sites.

2)   Energy transfer mechanism which is only operable up to an absorbed dose of about 0.5×10^{21} eV/g.

3) Radiation induced back reaction:

$$NO_2^- + \tfrac{1}{2} O_2 \xrightarrow{\text{-\hspace{-2pt}\raisebox{0pt}{$\sim\!\!\sim$}\hspace{-2pt}}} NO_3^-;$$

this reaction is distinguished from reaction (3-4) in which the initial oxygen fragment reacts with the parent NO_2^- to form NO_3^-.

4) A possible reaction of NO_2^- with electron deficient centers or cation to give stable NO_2. To account for the presence of stable NO_2 in some nitrates.

5) Little or no decomposition of nitrite ion. All available evidence points to the fact that nitrite ion is stable in the nitrite lattice i.e. little or no decomposition of nitrite ion occurs during irradiation of nitrates.

References

50 V. V. Sviridov "Photochemistry and radiation chemistry of solid inorganic substances", *Izv. Vys. Shkola*, Minsk (1964)

51 Ym. A. Zakharov and V. A. Nevostruev, *Russian Chemical Reviews* (1) (1968) U.D.C. 541, 281

52 L. K. Narayanswamy, *Trans. Far. Soc.* **31**, 1411 (1935)

53 A. O. Allen and J. A. Ghormley, *J. Chem. Phys.* **15**, 208 (1947)

54 R. D. Smith and A. H. W. Aten, *J. Inorg. and Nucl. Chem.* **1**, 296 (1955)

55 C. J. Hochanadel and T. W. Davis, *J. Chem. Phys.* **27**, 333 (1957)

56 J. Cunningham and H. G. Heal, *Trans. Far. Soc.* **54**, 1355 (1957)

57 H. G. Heal, *Can. J. Chem.* **31**, 91 (1953)

58 E. R. Johnson, *J.A.C.S.* **80**, 4460 (1958)

59 J. Forten and E. R. Johnson, *J. Phys. and Chem. of Solids* **15**, 218 (1960)

60 A. S. Baberkin, *Russ. Jour. of Phys. Chem.* **35**, no. 2, 179 (1961)

61 D. Hall and G. N. Walton, *J. Inorg. and Nucl. Chem.* **6**, 288 (1958)

62 D. Hall and G. N. Walton, *J. Inorg. and Nucl. Chem.* **10**, 215 (1959)

63 J. Cunningham, *J. Phys. Chem.* **65**, 628 (1961)

64 E. R. Johnson and J. Forten, *Disc. of Far. Soc.* No. **31**, 238 (1961)

65 A. R. Jones and R. L. Durfee, *Rad. Res.* **15**, 546 (1961)

66 A. R. Jones, *Chemistry Progress Report*, ORNL, period ending June 20 (1961)

67 E. R. Johnson, *J. Phys. Chem.* **66**, 755 (1962)

68 C. J. Hochanadel, *Rad. Res. Vol.* **16**, No. 3, 286 (1962)

69 T. H. Chen and E. R. Johnson, *J. Phys. Chem.* **66**, 2068 (1962)

70   J. Cunningham and L. R. Steele, *Phys. Rev. L.* **9** No. 2, 47 (1962)

71   S. R. Logan and W. J. Moore, *J. Phys. Chem.* **67**, 1042 (1963)

71a C. H. Hsiung and A. A. Gordus, *J. Chem. Phys.* **36**, 947 (1962)

72   J. Cunningham, *J. Phys. Chem.* **67**, 1772 (1963)

73   J. Cunningham, *J.A.C.S.* **85**, 2716 (1963)

74   J. Cunningham, *Trans. Far. Soc.* No. **525**, 62, 2423 (1965)

74a V. V. Boldyrev, V. M. Lykhin, A. N. Oblivantsev and K. M. Salikhov, *Kinet. i Katal.* **7**, 932 (1966)

75   E. J. Henley and E. R. Johnson, *Chemistry and Physics of High Energy Reaction* p. 253–261, Washington, D. C. University Press (1969)

76   J. Cunningham, *J. Phys. Chem.* **70**, 30 (1966)

77   T. G. Ward, G. E. Boyd and R. C. Axtman, *Rad. Res.* **33**, 447 (1967)

78   M. O. Kostin, R. C. Axtman and E. F. Johnson, "Decomposition of molten lithium nitrate" *USAEC Report* MATT-329 (1965)

79   S. H. Cho and E. F. Johnson, Kinetics of the thermal decomposition of lithium nitrate", *USAEC Report* MATT-324 (1964)

80   T. G. Ward, E. C. Axtman and G. E. Boyd, *Rad. Res.* **33**, 456 (1968)

81   F. Vratny and G. Gugliotta, *J. Inorg. and Nucl. Chem.* **20**, 252 (1961)

82   F. T. Jones, Chemistry Department, Stevens Institute of Technology, Hoboken, N. J., *private communication*

82a S. Kaucic and A. G. Maddock, *Trans. Far. Soc.*, **65**, 1083, 1969

83   P. Pringsheim, *J. Chem. Phys.* **23**, 369 (1955)

84   J. Cunningham, *Int. Journ. of Phys. and Chem. of Solids* **23**, 843 (1962)

85   E. Hyatchinson and P. Pringsheim, *J. Chem. Phys.* **23**, 1113 (1955) (this paper gives references to earlier work in these systems).

86   M. C. R. Symons, "ESR of radiation damage in Solids" *Radiation Chemistry* Vol. II (Advances in Chemistry Series 82) American Chemical Soc. Washington, D. C. (1968)

87   J. Cunningham in *Radical Ions* E. Kaiser and L. Kevan Eds. Wiley Inter-science, New York (1968)

88   W. B. Ard, *J. Chem. Phys.* **23**, 1967 (1955)

89   B. Bleaney, W. Hayes, P. M. Llewellyn, *Nature* **179**, 140 (1957)

90   H. Zeldes and R. Livingston, *J. Chem. Phys.* **35**, 563 (1961)

91   C. Jacard, *Phys. Rev. 124*, No. 1, **60** (1961)

92   J. Cunningham, *J. Phys. Chem.* **66**, 770 (1962)

93   H. Zeldes, *Paramagnetic Resonance* Vol. II p. 764 Academic Press Inc. New York (1963)

94   H. Zeldes and R. Livingston, *J. Chem. Phys.* **37**, 3017 (1962)

95   R. M. Golding and M. Henchman, *J. Chem. Phys.* **40**, 1554 (1964)

96   J. Cunningham, J. A. McMillan, B. Smaller and E. Yasaitis, *Int. J. of the phys. and Chem. of Solids* **23**, 167 (1962)

97   K. Zdansky and Z. Sroubek, *Czech J. Phys.* **1314**, 121 (1964)

98   K. Zdansky and Z. Sroubek, *Phys. Stat. Solidi* **7**, 167 (1964)

99   D. Shoemaker and E. Boesman, *Compt. Rend.* **252**, 2099, 2865 (1961)

100   R. Adde and P. Petit, *Compt. Rend.* **256**, 4682 (1963)

101   K. Gesi and Y. Kazumata, *J. Phys. Soc. Japan* **19**, 1981 (1964)

102   W. C. Mosely and W. G. Moulton, *J. Chem. Phys.* **43**, 1207 (1965)

103   J. Cunningham, *J. Phys. Chem.* **71**, 1967 (1967)

104   R. W. Holmberg and R. Livingston, *J. Chem. Phys.* **47**, 2552 (1967)

4

Radiation induced decomposition of the halates

Studies on the radiation induced decomposition of the halates have revealed the complex nature of the decomposition. Experiments have established the occurence of complex thermal and radiation induced reactions of the products. ESR and optical spectra have revealed the presence of several radicals and radical ions that are stable at room temperature and whose role in the kinetics of the decomposition is yet to be determined. As will become apparent there remains much to be done before our understanding of these compounds is in a satisfactory state.

4.1 Bromates

Many of the studies on the bromates are those associated with enrichment (retention) of radioactive bromine using a Szilard–Chatmers separation[105,106]; only those studies bearing some direct relation to the kinetics of the decomposition will be discussed in what follows.

Boyd and Cobble[107] studied the relative contributions of slow and fast neutrons, gamma rays and beta rays to the observed decomposition of crystalline $KBrO_3$. They made no attempt to identify the products of the decomposition or to study the kinetics and only reported on gross decomposition (bromine, bromide and hypobromide). They observed a G value for the decomposition using gamma rays of 1.0 ± 0.1. A similar G value was estimated for the decomposition using 1.15 MeV electrons. The G values estimated for fast neutrons (due to "knock on" collisions with subsequent recoil) was 23–30.

Maddock and Muller[108] studied retention of bromine activity in calcium bromate irradiated with slow neutrons. No attempt was made in these studies to determine the products of the decomposition. Bromine retention in the form of water soluble bromide, hypobromide, etc. only was investigated.

Boyd, Graham and Larsen[109] reported on a rather extensive study of the decomposition of the alkali bromates using cobalt-60 gamma rays. They analyzed for bromite, hypobromite, oxygen and bromite ion; in addition, they also determined "total" oxidizing power. Irradiations were done at different temperatures in conjunction with studies on the thermal decomposition.

The irradiated salts had a light yellow to gold color depending on the absorbed dose, which was readily bleached by gentle heating; the disappearence of the color being accompanied by a simultaneous loss in oxidizing power. Initial G values for the decomposition of the various salts are summarized in Table 4-1 below. No effect of crystal size (surface area) or

TABLE 4-1   Decomposition yields for the alkali metal bromates

Salt	Dose (eV mole^{-1} × 10^{-23})	Radiolytic yields (molec./100 eV)		
		"G(Ox.)"	G(Br$^-$)	$-G$(BrO$_3^-$)
LiBrO$_3$	6.44	0.21	0.13	0.33
NaBrO$_3$	0.491	0.87	0.65	1.52
NaBrO$_3$	5.30	0.55	0.98	1.52
NaBrO$_3$	6.06	0.64	0.88	1.48
KBrO$_3$	0.547	0.87	0.83	1.61
KBrO$_3$	5.90	0.46	1.07	1.53
KBrO$_3$	6.56	0.41	0.97	1.35
RbBrO$_3$	9.99	0.82	1.02	1.85
CsBrO$_3$	16.22	0.49	2.23	2.73

dose rate on the decomposition was observed. The temperatures dependence of the radiation induced decomposition indicated a small change over an interval of 280°C. These results are summarized in Table 4-2 below.

The thermal decomposition of lithium, cesium and potassium bromates was studied in temperature range of 150–350°C. The decomposition was found to be linear with time and Arrhenius plots for the three salts were parallel and gave an activation energy of 40 ± 2 kcal per mole. It was noted that this is the value of the bromine-oxygen bond. The thermal

TABLE 4-2 Temperature dependence of radiolysis of alkali metal bromates

	Temp. (°C)	Dose, (eV mole$^{-1} \times 10^{-23}$)	Decomposition mole Br$^-$/mole BrO$_3^-$	Molec./ 100 eV
LiBrO$_3$	85	0.753	0.408	0.33
	−195	0.753	0.327	0.26
NaBrO$_3$	85	0.150	0.361	1.45
	−195	0.150	0.251	1.01
NaBrO$_3$	85	0.491	1.235	1.52
	−195	0.491	0.980	1.20
KBrO$_3$	85	0.167	0.476	1.72
	−195	0.167	0.368	1.33
KBrO$_3$	85	0.457	1.360	1.79
	−195	0.457	1.064	1.40
KBrO$_3$	85	0.547	1.465	1.61
	−195	0.547	1.313	1.45
KBrO$_3$	85	0.615	1.628	1.59
	−195	0.615	1.490	1.45
KBrO$_3$	85	0.947	2.560	1.63
	−195	0.947	1.883	1.20
RbBrO$_3$	85	0.762	2.700	2.14
	−195	0.762 a	1.837	1.45
CsBrO$_3$	75	0.903	5.424	3.62
	−195	0.993	4.678	3.12
Ba(BrO$_3$)$_2$	95	0.373	1.188	1.92
	−195	0.373	1.063	1.72

decomposition rates were LiBrO$_3$ > KBrO$_3$ > CsBrO$_3$, which is the reverse of the radiation induced decomposition.

The authors found no correlation between radiolytic yields, free energies of formation or decomposition (to bromide ion plus oxygen gas) but did find a correlation with free space. Bromite, hypobromite and bromide ion (plus oxygen) were the only stable species detected. A reaction mechanism was developed in which the following reactions were thought to be the most significant:

$$\text{BrO}_3^- \leadsto \text{BrO}_3^-* \qquad (4\text{-}1)$$

$$\leadsto \text{BrO}_3 + e^-. \qquad (4\text{-}2)$$

$$\text{BrO}_3^-* \longrightarrow \text{BrO}_2^- + O \qquad (4\text{-}3)$$

$$\longrightarrow \text{BrO}^- + O_2 \qquad (4\text{-}4)$$

$$\longrightarrow \text{Br}^- + O + O_2 \qquad (4\text{-}5)$$

$$O + O \longrightarrow O_2 \qquad (4\text{-}6)$$

The authors felt that since the abstraction reaction

$$BrO_3^- + O \longrightarrow BrO_2^- + O_2 \qquad\qquad (4\text{-}7)$$

required appreciable thermal activation, it was of minor importance in the decomposition.

Anderson[110] observed a marked effect of pressure on the thermal annealing of irradiated $KBrO_3$. Heating the irradiated sample at 135°C for 10 minutes under 2000 atm. pressure produced $\sim 60\%$ decrease in total oxidizing fragments and an 11% decrease in bromide content compared to 8% decrease in oxidizing fragments and zero % decrease in bromide for a sample heated at this temperature at 1 atm. for 10 minutes. Annealing of radiolytic damage occurred at room temperature under compression. A 20% decrease in oxidizing power was observed for a sample compressed at 2000 atm. at 20°C.

Boyd and Larson[111] studied the decomposition of the alkali metal bromates using neutrons from the Oak Ridge graphite reactor as the source of radiation. Dosimetry was done using aqueous ceric sulfate solutions calibrated calorimetrically. Great care was used in determining total absorbed dose from the various reactor radiations and emanations from the induced radioactivity. These included reactor gamma ray dose, capture gamma rays, recoils, decay radiation, and neutron scattering dose. The products of the decomposition were believed to be only BrO_2^-, BrO^-, Br^- and O_2. The decomposition yields were not linear with dose, as have been found using only gamma rays as the radiation source;[109] indicating some back reaction was occurring. With the exception of $LiBrO_3$ the initial yields were identical with those reported previously (see Table 4-3). More than 90% of the absorbed dose could be attributed to pile gammas, capture gammas and decay radiations hence this favorable comparison with the gamma rays experiments is not too surprising. With $LiBrO_3$ on the other hand about 95% of the absorbed dose was due to recoils from $^6Li(n, \alpha)^3H$ reaction.

A more careful study of the LET dependence on $LiBrO_3$ decomposition was made[112] using $LiBrO_3$ containing 0.01 to 1.0 atom-% ^6Li. The radiation dose was estimated using the relation.

$$D(\text{eV mole}^{-1}) = \phi t\, \sigma NE\, \theta fc \qquad\qquad (4\text{-}8)$$

where

$\quad \phi \quad$ thermal neutron flux

$\quad t \quad =$ time

$\quad \sigma \quad =$ effective neutron capture cross section of ^6Li

TABLE 4-3 Summary of data for pile irradiation of alkali bromates

Salt	γ-Ray dose	γ-Capture rays	Recoils	radiation	Energetic neutron scattering dose	Total dose	Decomposition mmole⁻¹ per neutron cm⁻³ × 10¹⁶	Gross reactor yield $G_0(-BrO_3^-)$	Cobalt-60 gamma-ray yield $G_0(-BrO_3^-)$
$LiBrO_3$	2.25	1.41	0.223	1.72	0.18	5.78	0.437	0.47	0.31
$NaBrO_3$	2.46	1.47	0.002	1.66	0.15	5.74	1.367	1.4	1.42
$KBrO_3$	2.72	1.76	0.002	1.65	0.13	6.26	1.279	1.3	1.32
$RbBrO_3$	3.46	1.43	0.002	1.67	0.15	6.78	2.694	2.4	2.33
$CsBrO_3$	5.37	7.77	0.002	1.54	0.17	14.85	7.570	3.1	3.4
$LiBrO_3$	2.25	1.14	83.42	1.39	0.18	88.38	21.198	1.4	0.31

Dose values in MeV. mole⁻¹ per neutron cm⁻².

N = Avogadro's number
E = Energy released per fission (4.787 MeV)
θ = Fraction of recoil energy absorbed (assumed to be unity)
f = Self-shielding factor (6.8 % at maximum ^{6}Li concentration).
c = Atomic % of Lithium-6

   Samples containing varying amounts of ^{6}Li were exposed to pile radiation for 2, 4, and 6 hrs. Total oxidizing power (oxidizing fragments) and bromide were determined. For the 2 hour exposure the decomposition was linear

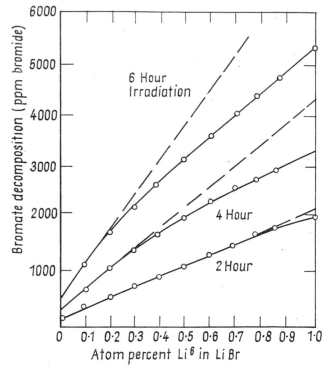

FIGURE 4-1   Dependance of bromate radiolysis on Li^6 concentration in LiBrO$_4$ and on irradiation time

with ^{6}Li content indicating that the radiolysis was independent of dose rate (see Figure 4-1). The yields of oxidizing power (presumably BrO$_2^-$ and BrO$^-$) or Br$^-$ when plotted against dose fall on a single line and shows that there is no change in yield of these species with dose rate (Figure 4-2, 4-3). The initial oxidizing yield $G("ox")$, the initial Br yield, $G_0(Br^-)$ and total bromate

$G_0(-BrO_3^-)$ were determined to be 1.1, 0.38 and 1.48 respectively. The corresponding initial G value using cobalt-60 gamma radiation (LET = 0.06 eV/Å) were 0.21, 0.10 and 0.31 respectively. The ratio G(pile)/G (gamma) for oxidizing fragments, Br^- and total bromate decomposition were 5.2, 3.8 and 4.8 indicating very little difference in product distribution with LET. It was observed that most of the oxidizing species could be

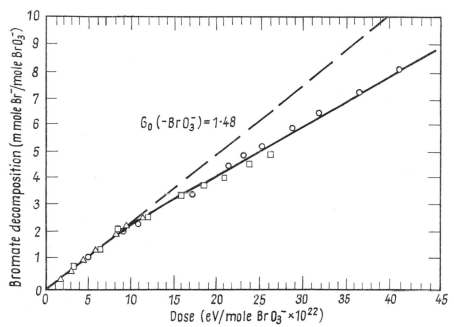

FIGURE 4-2 Radiolysis of crystalline LiBrO₃ by ⁶Li fission recoil particles. △ = 2 hour irradiation, □ = 4 hour irradiation, ○ = 6 hour irradiation.

removed by thermal annealing at relatively low temperatures hence a thermal spike hypothesis, as being responsible for the increase in initial G values with increased LET, was rejected.

Ramasastry and Murti[113] investigated the optical absorption (UV) and electron spin resonance spectra of irradiated sodium bromate. The irradiated crystals showed no paramagnetic resonance spectra but did reveal three absorption bands in the UV at 280, 330 and 420 mμ. The band at 330 mμ was attributed to the hypobromite ion. The band at 280 mμ was interpreted as a charge transfer peak of the bromate ion influenced by the presence of hypobromite ions in its vincinity. The band at 420 mμ was tentatively assigned to Br_2.

A more detailed study of bromate radiolysis was undertaken by Boyd et al.[114] using $CsBrO_3$ as the principal salt because of its ease of decomposition. The decomposition was studied as a function of dose rate, total dose and temperature. Cobalt-60 gammas was used as the radiation source and absorbed dose was determined using Fricke dosimetry ($G = 15.6$) with corrections for the differences in mass absorption coefficients for the salt

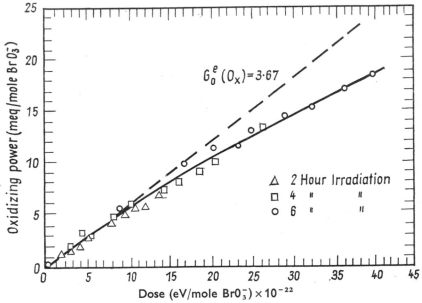

FIGURE 4-3 Production of oxidizing fragments in the radiolysis of crystalline $LiBrO_3$ by ^6Li fission recoil particles. $\triangle$ = 2 hour irradiation, $\square$ = 4 hour irradiation, $\bigcirc$ = 6 hour irradiation

and dosimeter solution. Chemical analysis of an aqueous solution of the irradiated salt revealed the presence of BrO_2^-, BrO^-, Br^- and O_2. All tests for free bromine were negative. Kinetic analysis of data did however, indicate an additional oxidizing specie (other than BrO_2), present in small amount which the authors believed to be BrO_3 or BrO_4^-. The decomposition yield of bromate ion as a function of absorbed dose showed an initial increase and then became linear beyond an absorbed dose of about 5×10^{23} eV/mole. The yield of oxygen was linear with dose while that of the oxidizing fragment was initially dose dependent and then appeared to saturate. In contrast, the yield of bromide ion was small initially but increased with absorbed dose (see Figure 4-4). Initial G values at 35°C were $G(-BrO_3^-) = 5.5$; $G(Br^-) = 0.55$; $G(O_2 = 2.4)$; and $G("oxidizing") = 5.0$. The decomposition appeared to be dose rate independent for a 10 fold variation in dose rate, however, there was a very definite temperature dependence of the oxidizing fragment yield and total bromate decomposition (Table 4-4).

TABLE 4-4 Temperature dependence of the yields for $CsBrO_3$ decomposition

Temp., °C	Oxidizing fragments (mequiv. mole^{-1})	bromate decompn. (mmoles mole^{-1})
−196	53.2	14.1
−86	57.8	15.6
40	45.9	15.8
100	30.0	10.2
200	11.0	15.5
297	0.0	40.0

Dose of 2.6×10^{23} eV mole^{-1}

Post-irradiation annealing studies established that the oxidizing fragments undergo at least two reactions; one reaction to form bromide ion and another to give bromate ion. The results therefore show that a thermally induced back reaction of decomposition products takes place and provides a good explanation for the fact that bromate decomposition is greater at −78°C than 57°C.

The decomposition of the alkaline earth bromates[115] was undertaken to further illucidate the role of crystal environment in the radiolysis of the bromates. The salts studied were magnesium, barium, calcium and strontium bromates. It was observed that in the preparation of the anhydrous salts, $Mg(BrO_3)_2$ underwent extensive thermal decomposition (105°C) and $Ba(BrO_3)$ appeared to undergo one or more phase transitions. Consequently initial G values were only considered valid for the calcium and strontium salts (Table 4-5). All irradiations were done with cobalt-60 gammas. Ferrous sulfate dosimetry was used with $G(Fe^{3+}) = 15.6$.

The extent of decomposition of $Ba(BrO_3)_2$ per unit of absorbed energy varied with drying temperature. The order or stability for alkaline earth bromate was Mg > Ca > Sr > Ba. The decomposition of bromate ion in

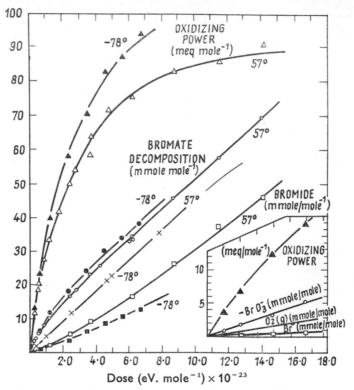

FIGURE 4-4   Radiolysis of crystalline CsBrO$_3$ with ^{60}Co γ-rays from source a. Insert shows data for radiolysis at low dose with source b.

these salts increased nonlinearly with absorbed dose at low conversions and then approached linearity. The rate at which bromide ion formed appeared to increase linearly. The rate at which bromide ion formed appeared to increase with increasing dose suggesting that part of the bromide ion arose from the thermal or radiation induced decomposition of some intermediates.

TABLE 4-5   Initial G values for radiolysis of alkaline earth bromates

	Ca(BrO$_3$)$_2$	Sr(BrO$_3$)$_2$
G($-$BrO$_3$)	2.36	2.87
G(Br$^-$)	0.33	0.51
G("ox")	2.03	2.36
G(O$_2$)	1.51	1.96

Products identified as "oxidizing fragments" also were dose dependent and appeared to saturate at large doses as was found with $CsBrO_3$ radiolysis. The stoichiometric composition of an average "oxidizing fragment" was computed from the difference between the amount of $O_2(g)$ expected from the amount of BrO_3^- decomposed and that actually observed experimentally. This difference was divided by the amount of bromine in the fragments and an average composition of a fragment was calculated as a function of absorbed dose. The composition of an average fragment in $Ca(BrO_3)_2$ decreased from $BrO_{2.0}$ to $BrO_{1.6}$ and in $Sr(BrO_3)_2$ it decreased from $BrO_{2.0}$ to $BrO_{1.63}$. If the initial composition of the oxidizing fragments had been only BrO_2^- and BrO^- then the average composition would have been less than $BrO_{2.0}$. The oxidizing fragments could not consist solely of BrO_2^- because the fragments had an average oxidizing number* of 3.5. As a result of these analyses the authors concluded that the composition of the oxidizing species were BrO_2^-, BrO_2, and BrO^-. Initial yields of BrO_2 were small and apparently this specie decomposed (reacted or escaped from the lattice) readily since it did not exist beyond an absorbed dose of 1×10^{23} eV/mole. A direct identification of this specie was not obtained. A mechanism for the decomposition was given which postulated that BrO_2 was the precursor for the observed hypobromite ion

$$BrO_3^- \overset{\sim\!\!\wedge\!\!\sim}{\longrightarrow} BrO_2^- + O \qquad (4\text{-}9)$$

$$\overset{\sim\!\!\wedge\!\!\sim}{\longrightarrow} Br^- + 3\,O \qquad (4\text{-}10)$$

$$\overset{\sim\!\!\wedge\!\!\sim}{\longrightarrow} BrO_2 + O + e \qquad (4\text{-}11)$$

$$BrO_2 + e \longrightarrow BrO_2^- \qquad (4\text{-}12)$$

$$\longrightarrow BrO^- + O \qquad (4\text{-}13)$$

$$BrO^- \longrightarrow Br^- + O \qquad (4\text{-}14)$$

$$BrO_2^- + O \longrightarrow BrO_3^-. \qquad (4\text{-}15)$$

A subsequent paper by these authors[116] given at the ASTM symposium on "Chemical and Physical Effects of High Energy Radiation on Inorganic Substances" is essentially a summary of previous work.

Herley and Levy[118] investigated the photochemical decomposition of sodium bromate using a GE BH 6 high pressure mercury lamp. Decomposition was measured by determining the oxygen pressure over the photolyzed

---

* The oxidizing number is the ratio of the total oxidizing power in millieq. per mole to the mmoles per gram of bromine contained in the fragments.

sample. The decomposition showed an initial rapid deceleratary period which slowly leveled off to a steady rate after about 100 min. exposure. Preirradiating the samples with cobalt-60 gamma rays had the effect of decreasing the initial rate and the time to reach the steady state rate. The steady state rate fit an expression of the type

$$R(I) = \frac{\alpha I^2}{(1 \pm \beta I)} \qquad (4\text{-}16)$$

where $R(I)$ = rate of decomposition
I = Intensity
α, β = Constants.

To explain the decomposition versus time curves the authors postulated that decomposition occurs via excited species at certain lattice sites and at least part of the time these sites are doubly excited.

Andersen, Madsen und Olesen[117] in their studies on the annealing of irradiated $KBrO_3$ present evidence for the formation of species other than BrO^-, BrO_2^- and Br^-. Samples irradiated at $-78°$ with 10 MeV electrons to an absorbed dose of 5.22 eV/mole showed the presence of BrO^-, BrO_2^-, Br^- plus oxygen. The G values for the bromine fragments were 0.27, 0.59, and 0.33 respectively. Samples exposed in the thermal column at $-78°C$ gave only $^{82}Br^-$, $^{82}BrO_3^-$ and a small amount of $^{82}Br_2$ (3%) as the only recoil products. The UV spectra of $NaBrO_3$ irradiated at room temperature shows three optical absorption bands at 280, 330, and 340 mμ and were attributed to a BrO_3^- ion influenced by the presence of a BrO^- ion, BrO and Br_2. Aqueous solutions of the irradiated salt when extracted with CCl_4 show a band at 410 mμ in the CCl_4 fraction. Several paramagnetic centers were observed in irradiated $NaBrO_3$ but were not identified. The authors postulate the presence of charge centers such as $(BrO–BrO_3)^-$, O_2^-, O_3^-, $(Br–BrO_3)^-$ and BrO_3^{2-} to account for the observed annealing kinetics.

Andersen, Byberg and Olsen[119] have observed the presence of a number of paramagnetic centers in irradiated $NaBrO_3$. Heating the irradiated crystals to 80–100°C for several days destroys most of the centers. One center which they believe due to O_3^- ion appears to be relatively stable and is not destroyed unless the annealing temperature is greater than 100°C. The principal g values for this center when compared to those for O_3^- observed in other compounds (see Table 4-6) appears to be slightly higher.

Gamble[120] examined the ESR spectrum of single crystals of $NaBrO_3$ irradiated at room temperature with cobalt-60 gamma rays. The spectrum was examined at room temperature and at 77°K. The spectrum revealed the presence of spin $\frac{1}{2}$ centers superimposed upon a broader asymetric resonance. The spin $\frac{1}{2}$ centers were located at four nonequivalent sites having axial symmetry with the symmetry axis of each site along a [111] direction of the crystal. The measured values for the centers were $g_{11} = 2.0038$ and $g_{\perp} = 2.0218$. The center was tentatively assigned to the O_3^- ion. The g values found by Gamble are in excellent agreement with those of reference (119) for this center (see Table 4-6).

TABLE 4-6   Principal g values observed for O_3^-

Species	Matrix	Gxx	Gyy	Gzz	Gav	Ref.
O_3^-	$KClO_4$	2.0174	2.0113	2.0025	2.0104	a
O_3^-	$KClO_4$	2.018	2.011	2.0034	2.011	b
O_3^-	NaO_3				2.012	c
O_3^-	NaO_3	2.015	2.015	2.003	2.011	d
	$NaBrO_3$	2.022	2.022	2.006	2.017	e

(a) P. W. Atkins, J. A. Brivati, N. Keen, M. C. R. Symons, and P. A. Trevalion, *J. Chem. Soc.*, **4785** (1962). (b) A. V. Dubovitskii and G. B. Manelis, *Kinetika i Kataliz*, **6**, 828 (1965). (c) A. D. McLachlan, M. C. R. Symons, and M. G. Townsend, *J. Chem. Soc.*, **952** (1959). (d) J. E. Bennett, D. J. E. Ingram, and D. Schonland, *Proc. Phys. Soc.* (*London*), *A***69**, 556 (1965). (e) reference 119.

A study of ^{82}Br recoils in neutron irradiated alkali metal bromates indicates that thermal reactions of the irradiated salts are indeed complex.[121] Radiobromine was observed in the oxidation states BrO_3^-, BrO_2^-, BrO^- and Br^-. The fraction of ^{82}Br retained as BrO_3^- was about the same for all the alkali bromates but there were large differences in the relative yields of BrO_2^- and BrO^-. Thermal annealing produced an increase in radio-bromate and a related decrease in radiobromide. Ninety percent of the radiobromide in neutron irradiated $LiBrO_3$ was oxidized to radiobromate after annealing for 6 hrs. at 200°C. A very definite increase in radiobromate was observed when neutron irradiated bromates were irradiated with cobalt-60 gammas at temperature 0°C and higher. No change in radio-bromate in neutron irradiated bromate was observed when irradiations were done at -78°C. Although some neutral bromine could be extracted

7*

from aqueous solutions this was thought to arise from BrO_2^- and not from any Br_2 or Br in the irradiated salt.

In a continuing effort to identify the stable species produced in the decomposition of the bromates Boyd et al.[122] observed the formation of the perbromate ion in gamma irradiated $CsBrO_3$. Appelman[123] had reported on the preparation of $RbBrO_4$ by oxidation of bromate ion in aqueous solution with XeF_2. Boyd et al. determined the IR and Raman spectra of $RbBrO_4$ so prepared and used this spectra in the identification of $CsBrO_4$. The infrared spectrum of irradiated $CsBrO_3$ (total absorbed dose = 2.9 $\times 10^{24}$ eV/mole) was obtained using a KBr disk containing 2.4 wt. % of the irradiated (or unirradiated) salt. The IR spectrum revealed the presence of BrO_2^- in addition to BrO_3^- and BrO_4^-. The relative order and spacing of the bands associated with these ions were consistent with those for the analogous chlorine oxyanions. Table 4-7 gives the vibrational frequencies of the perhalate ions. The G value for BrO_4^- fromation for an absorbed dose of 3.2×10^{24} eV/mole was 0.16 whereas that for $G(Br^- + BrO^- + BrO_2^-) = 1.21$ and for BrO_3^- disappearance it was 1.37.

TABLE 4-7   Vibrational frequencies of perhalate ions (cm^{-1})

Ion	$\nu_1(A_1)$	$\nu_2(E)$	$\nu_3(F_2)$	$\nu_4(F_2)$
ClO_4^-	935	460	1110	630
BrO_4^-	801	331	878	410
IO_4^-	791	256	853	325

No BrO_4^- was detected in neutron irradiated $^7LiBrO_3$. The authors placed an upper limit of less than 0.05% on radioperbromate in these experiments.

4.1.1   Summary of bromate radiolysis

As is apparent from what has preceeded the bulk of the research done on the bromate has been by Boyd and his associates. Consequently there is very little controversial data.

4.1.1.1   Primary products   EPR spectra of irradiated bromates indicates the presence of several thermally (up to 50°C) stable paramagnetic species. A tentative identification of only one of these species has been made and that to O_3^-. The assignment of the 420 mµ band to the presence of Br_2 by

Ramasastry[113] does not appear to be valid especially since the O_3^- has been observed in other halates in the wave length region 420–450 mμ. The stable products from the room temperature radiation induced decomposition of bromates appear to be O_2, Br^-, BrO_2^-, and BrO_4^-. It is presumed that BrO_4^- is not a primary product but arises from a radiation induced reaction of the products. BrO^- and BrO_2^- have been identified in aqueous solution (UV spectrum), BrO_2^- has been identified from IR spectra of the irradiated solid as has BrO_4^-. Ramasastry and Murti have attributed an absorbtion band at 330 mμ to the BrO^- ion however, the experimental evidence on $CsBrO_3$ decomposition supports the view that BrO^- is not a primary product but arises from a radiation induced reaction of BrO_2. BrO_2 has been determined to be an initial product of the radiolysis by a kinetic analysis of the data, however, it has not been identified by any direct method. In view of the direct observation of BrO_4^- the evidence for BrO_2 is less certain. Bromine has not been observed as a product in the decomposition of the bromates. Initial G values for the decomposition of the bromates are summarized in Table 4-8.

TABLE 4-8   Initial G values for radiolytic products of the crystalline alkali metal and alkaline earth bromates at 35°C

	$LiBrO_3$	$NaBrO_3$	$KBrO_3$	$RbBrO_3$	$CsBrO_3$	$Ca(BrO_3)_2$	$Sr(BrO_3)_2$
$G_0(-BrO_3^-)$	0.31	1.4	1.3	2.3	5.6	2.4	2.9
$G_0(Br^-)$	0.10	(0.59)	0.63	(0.96)	0.48	0.33	0.51
$G_0("Ox")^a$	0.21	0.83	0.69	1.4	5.2	2.0	2.4
$G_0(O_2)$	—	—	—	—	2.4	1.5	2.0
$G(BrO_4^-)$	—	—	—	—	0.16	—	—

^{a} $G("Ox")$ is presumed to mean $G(BrO^- + BrO_2^- + O_2)$.

4.1.1.2   Kinetics   The concentration of the oxidizing species increases non linearly and appears to reach a steady state. This effect is apparently caused by thermal and radiation induced reactions to form BrO_3^- and Br^-. Initial yields of bromide ion appear to be linear but with increasing absorbed dose the yields became progressively larger. Oxygen yields were approximately linear with absorbed dose up to 1.5×10^{24} eV/mole, however O_2 from $CsBrO_3$ decomposition showed a slight non-linear dependence at small absorbed doses.

4.1.1.3 Radiation and thermal induced back reaction A radiation induced reaction of the products to give BrO_3^-, Br^- and BrO_4^- appears to occur. Thermal annealing also induces a complex reaction of the products to give Br^- and BrO_3^- but not BrO_4^-.

4.2   Chlorates

The Chlorates have been more thoroughly investigated than bromates. In addition to determinations of products and product distribution there has been considerable effort in ESR and optical spectra. We will first discuss those papers dealing with the kinetics of the decomposition, this will be followed by discussions of the results of ESR and optical spectra and a final summary.

McCallum and Holmes[124] and Sharman and McCallum[125] investigated the chemical effects following the $^{35}Cl(n, \alpha)^{36}Cl$ reactions in crystalline sodium chlorate. In the latter reaction the authors observed that the major decomposition products (due largely to the intense gamma radiation to which the samples were exposed) were Cl^-, ClO_4^- and small amounts of the chlorine compounds of intermediate valence such as ClO_2^-, ClO_2, ClO^- (none of which was positively identified).

Baberken[126] studied the decomposition of $KClO_3$ at low absorbed doses of cobalt-60 gamma rays. The products observed were ClO_2^-, Cl^- and O_2. Heating the irradiated samples to 200°C caused a 30% decrease in the ClO_2^- $(G = 1.2)$ concentration. The chlorite ion was presumed to decompose to Cl^- plus O_2.

Heal[127] studied the decomposition of potassium chlorate using 50 kvp X-rays as the radiation source. Dosimetry was determined calorimetrically Irradiation were done at $-196°$ and 25°C and at dose rates varying between 0.12 and 1.24×10^{23} eV/mole hr. The irradiated samples were examined directly by UV and IR spectroscopy and in aqueous solution. The compounds detected with reasonable certainty were ClO^-, ClO_2, ClO_2^-, Cl^- and O_2. Aqueous solutions of irradiated $KClO_3$ showed an absorption maxima in the region 3000–4000 Å which gradually disappeared on standing. This absorption maxima was attributed to chlorine dioxide. The spectra of the irradiated solid at 25°C showed the presence of two maxima which were definitely assigned to ClO^- and ClO_2^-. The yields of both ClO^- and ClO_2^- were dose dependent and reached a steady state at relatively low absorbed doses. The rate of formation of Cl^- increased initially with increasing absorbed dose but then become constant. The rate of formation of oxygen appeared to be constant. The total amount of oxygen evolved was observed to be less than required by stroichiometry from the decomposition to give ClO^-, ClO_2^-, ClO_2, and Cl^- and so Heal postulated that this deficiency was most likely due to the formation of perchlorate ion. Infrared absorption measurements on irradiated crystals showed a small peak at $1113\ cm^{-1}$ (ClO_4^- maxima $= 1100\ cm^{-1}$), however, Heal believed that this was not convincing proof of the existence of ClO_4^-. The spectra of samples irradiated at 25° also revealed an absorption maxima at 4500 Å. When the irradiated crystal was cooled to $-196°C$ this band developed structure, which disappeared on warming to 25°C. Crystals irradiated at $-196°C$ and observed at this temperature had a different spectra from those irradiated at 25°C. The 4500 Å band was missing and there was absorption over a considerable range of wavelengths with one well defined maxima at 2400 Å. When these crystals were warmed to 25°C and then re-examined at $-196°C$ the 2400 Å band disappeared as did most of the other absorption; this was replaced by a spectrum very similar to that shown by samples that had been originally irradiated at 25°C. Heal does not give any tentative identification to the 2400 Å band but does assign the 4500 Å to Cl_2O_6. This band (4500 Å), however, has subsequently been identified[131] as being due to the O_3^- ion (this is discussed later under optical and ESR spectra). Optical absorption spectra of aqueous solutions of the irradiated salt revealed the presence of three bands which were identified as being due to ClO^- ion (2900 Å), ClO_2^- ion (2610 Å) and ClO_2 (3000 –4000 Å). G value for the principal species are shown in Table 4-9. The fact that the yields of ClO^- and ClO_2^-

TABLE 4-9   Initial G values for products in $KClO_3$ irradiation

Temp.	$G(ClO_4^-)$	$G(Cl^-)$	$G(ClO^-)$	$G(ClO_2^-)$	$G(O_2)$	$G(ClO_2)$
25°C	1.9 max	1.3	0.5	2.0	2.5	0.2
$-196°C$		1.1	0.5	1.5	2.0	

reached a steady state after prolonged exposure was believed to be due to a radiation induced decomposition of these ions. No thermal annealing experiments on the irradiated crystals were performed.

Heal postulated a rather simple mechanism wherein Cl^-, ClO^- and ClO_2^- are all formed from excited ClO_3^- ions; the mechanism is shown below:

$$ClO_3^- \longrightarrow Cl^- + O_2 + O \tag{4-17}$$

$$\longrightarrow ClO^- + O_2 \tag{4-18}$$

$$\longrightarrow ClO_2^- + O \tag{4-19}$$

$$ClO_3^- + O \longrightarrow ClO_4^-. \tag{4-20}$$

Patrick and McCallum[128] reported on the work of Burchill[129] who had studied the decomposition of sodium, potassium and barium chlorates. Burchill observed that oxygen, chloride ion and chlorite ion were the major products in solutions of the irradiated salts. He also identified ClO_2, ClO^- and ClO_4^- as products but in much smaller amounts.

Patrick and McCallum[128] confined their studies to the room temperature decomposition of sodium chlorate using cobalt-60 as the radiation source.

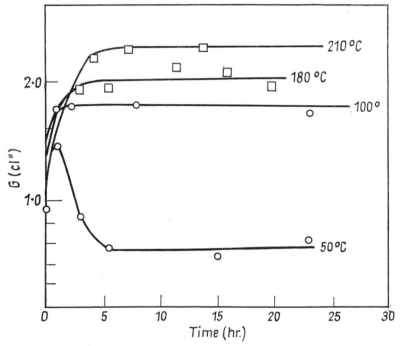

FIGURE 4-5   Change in $G(Cl^-)$ for irradiated $NaClO_3$ following thermal annealing for different times and temperature

Chemical analysis of aqueous solutions of the irradiated salt showed the presence of O_2, Cl^-, ClO^-, ClO_2^-, ClO_2 and ClO_4^-. The observed G values were 1.95, 0.95, 0.21, 1.34, 0.23 and 0.22 respectively. Annealing the irradiated crystals at 50°C produced an initial increase in $G(Cl^-)$ which reached a maximum at $G(Cl^-) = 1.38$. Continued heating at this temperature caused $G(Cl^-)$ to decrease to a steady state value of 0.55. Annealing at temperatures of 100° and above, however, caused an initial increase in $G(Cl^-)$ with no subsequent falling off i.e. $G(Cl^-)$ reaching a constant value (see Figure 4-5).

Burchill, Patrick and McCallum[130] in a continuation of previous studies on the chlorates investigated the radiation in induced decomposition of crystalline sodium, potassium and barium chlorates. The radiation source was cobalt-60 gammas and dosimetry was determined using ferrous oxidation with $G = 15.5$. Absorbed doses were corrected on the basis of compton absorption with a small correction for photoelectric absorption in the barium salt. The products, determined by analysis of aqueous solutions of the irradiated salts, were found to be O_2, ClO_2, ClO_2^-, ClO^-, ClO_4^- and Cl^- plus a small amount of H_2. The yield of products were found to increase linearly with absorbed dose and to be independent of dose rate over a 5 fold variation. The small amount of hydrogen observed was believed due to the action of trapped electrons with water. The perchlorate ion was believed due to the hydrolysis of Cl_2O_6. The observed G values for the products are shown in Table 4-10.

TABLE 4-10 G values for the radiation decomposition of chlorates

	$Ba(ClO_3)_2$	$KClO_3$	$NaClO_3$
ClO_2^-	2.05 ± 0.06	2.07 ± 0.07	1.32 ± 0.03
ClO^-	0.26 ± 0.06	0.27 ± 0.04	0.18 ± 0.03
ClO_2	0.25 ± 0.09	0.21 ± 0.04	0.13 ± 0.03
Cl^-	1.72 ± 0.07	1.64 ± 0.09	0.91 ± 0.03
ClO_4^-	0.60 ± 0.11	1.13 ± 0.06	0.22 ± 0.04
O_2	3.90 ± 0.04	2.76 ± 0.16	2.36 ± 0.25
H_2	0.11 ± 0.02	0.06 ± 0.03	0.05 ± 0.01

A study on the retention of radiochlorine[131] in irradiated $KClO_2$, $KClO_3$ and $KClO_4$ indicated that radiochlorine existed in the oxidation state previously observed in irradiated chlorates i.e. Cl^-, ClO_2^-, ClO_3^-, ClO_4^-. The products were determined by dissolution of the irradiated salt and anion-exchange chromatographic separation. The yield of ClO_4^- ion was very small. The decomposition via recoil of the chlorine atom showed significant difference in the relative yields of all products in the different salts.

A study of the decomposition of $KClO_3$[132] confirmed previous reports[39] on the existence of ClO_4^- ion in the lattice of the irradiated salt. Carefully dried samples were irradiated with cobalt-60 gammas. The dose rate as determined by Fricke dosimetry ($G = 15.6$) was 1.08×10^{20} eV/min. Absorbed dose in the $KClO_3$ was corrected using the ratio of the number of electrons per gram in $KClO_3$ to that in H_2O (dose $KClO_3 = 0.882$ dose H_2O). A calibration curve for ClO_4^- in $KClO_3$ was prepared by cocrystallizing varying amount of $KClO_4$ with $KClO_3$. The IR spectra of irradiated $KClO_3$ revealed the presence of ClO_2^- (806 and 844 cm^{-1}) and ClO_4^- (1108 cm^{-1}) Figure 4-6. The concentration of ClO_4^- was determined chemically and from the optical calibration curve. Agreement between the two methods was good especially at higher absorbed doses. The chemical method, however, was considered to be the more reliable. Since the agreement between the two methods was within $\pm 10\%$ this established that no appreciable amount of the ClO_4 could arise from hydrolysis of unstable specie trapped in the lattice. The G value for ClO_4^- appeared to decrease with increased absorbed dose. The initial G values for the products as determined by chemical analysis are shown in Table 4-11. Although ClO_2 was observed to be present it was not analyzed for.

TABLE 4-11 Yields of chlorine-containing species produced by ^{60}Co irradiation of crystalline $KClO_3$

Dose, eV (mol of $KClO_3$)$^{-1}$ $\times 10^{23}$	$G(Cl^-)$	$G(ClO^-)^a$	$G(ClO_2^-)$	$G(ClO_4)^*$
4.37	1.55 ± 0.05	0.32 ± 0.01	2.08 ± 0.03	0.97 ± 0.03
4.42	1.65 ± 0.05	0.32 ± 0.01	2.05 ± 0.03	0.94 ± 0.02
4.4	1.60 ± 0.07	0.32 ± 0.02	2.06 ± 0.04	0.96 ± 0.04
$\leqq 1.2$	1.64 ± 0.09	0.27 ± 0.04	2.07 ± 0.07	1.13 ± 0.06

[a] These data presented as $G(ClO^-)$ may contain a slight contribution from $G(ClO_2)$. Due to the strong $ClO_2(g)$ odor of the irradiated samples and to the close agreement between these values and Burchill, *et al.*,[130] values it is felt that the great majority of the ClO_2 escaped from the finely divided samples before dissolution for analysis.

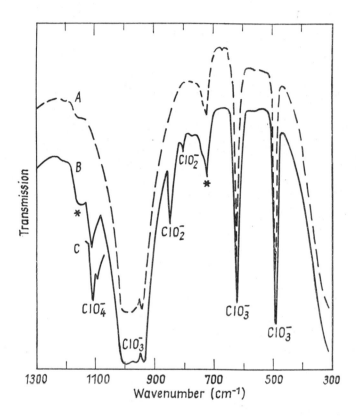

FIGURE 4-6   Infrared absorption by crystalline $KClO_3$ at room temperature (asterisks denote Nujol absorbancies) A = Unirradiated $KClO_3$, B = [60]Co γ-rays irradiated $KClO_3$, C = $KClO_4$—$KClO_3$ freeze-dried standard

4.2.1   Optical and paramagnetic resonance absorption spectra

Hasty, Ard and Moulton[133] examined the optical and paramagnetic resonance absorption spectra of single crystals of $KClO_3$ which had been irradiated with 50 kVp X-rays. The paramagnetic spectrum showed two sets of lines. One set was a single, relatively broad line with g value slightly greater than 2. The other set consisted of 16 equally intense lines with a line width of about 5 gauss. Examination of the spectra as a function of crystal orientation established that the 16 lines were four equally spaced groups of four equally spaced lines each. The separation between the lines was the same in a field of about 7000 gauss indicating the splitting was caused by hyperfine interaction. The authors attributed this center to a ClO radical which shares an electron with a neighboring ClO_3^- ion. The other center they attributed to O_2^-. Annealing the irradiated crystals at a temperature $>85°C$ caused the 16 line spectrum to disappear. No reference was made of the effect of annealing on the line attributed to O_2^-. The optical spectra of irradiated crystals revealed a single intense band at 4610 Å. This is apparently the same band observed by Heal[127] and as indicated previously is due to the O_3^- ion.

The electron spin resonance of $KClO_3$ irradiated at ambient temperatures and examined at $-196°C$ revealed the presence of several paramagnetic species.[134] One specie was definitely identified as the O_3^- ion and another believed to be a Cl atom strongly associated with a ClO_3^- ion. The authors do not believe that the radical designated by Hasty[133] *et al*. is the ClO radical but the Cl atom indicated above. The authors observed a 4600 Å absorption band in irradiated $KClO_4$ which they attribute to the O_3^- ion. The assignment of this band to the O_3^- ion was made on the basis that the band had structure similar to that found in ozonide ions in liquid ammonia.[135] Other paramagnetic centers were observed but could not be resolved.

Holmberg[136] observed a large number of paramagnetic centers in irradiated $KClO_3$ and $Ba(ClO_3)_2$ some of which showed hyperfine structure characteristic of chlorine nuclei. Only one center was resolved in $KClO_3$ and this was attributed to ClO_2. A similar specie was observed in irradiated $Ba(ClO_3)_2$ but the characterization was not definite.

Gamble[137] examined the paramagnetic resonance of single crystals of sodium chlorate which had been irradiated at room temperature with cobalt-60 gamma-rays. The centers were stable for several months at room temperature. At very low exposure doses (1.6×10^5 roentgens) a minimum of 3 paramagnetic centers was observed. Continued irradiation increased the density of these centers and in addition produced others. Only one spectra was resolved and this was attributed to either the ClO or the ClO_3 radical. The latter radical was favored based on two facts; one was that annealing destroyed the center but it could be restored by subsequent irradiation. It was felt that the ClO radical would not respond in this manner because the formation of ClO from ClO_3^- requires the loss of two oxygen atoms and a relocation of the Cl-O bond along a body diagnal of the cube.

The other fact was that the g values compared favorably with those of Cole[138] for ClO_3 found in irradiated NH_4ClO_4, however the hyperfine nuclear coupling for ^{35}Cl differed appreciably.

Subbotin[139] examined the paramagnetic centers in irradiated sodium, calcium, magnesium, strontium and barium chlorates. The spectra were observed at room temperature in samples irradiated at ambient temperatures with X-rays to high absorbed doses. Two paramagnetic centers were identified which were attributed to O_2^- and $(ClO-ClO_3^-)$ in agreement with those found by Hasty *et al.*[133] in $KClO_3$. The concentration of the centers was

TABLE 4-12   g factors for various chlorates

Substance	Paramagnetic center	g-factor value
$Mg(ClO_3)_2$	O_2^-	$g = 2.0318 \pm 0.0003$
	$ClO-ClO_3^-$	$g_{||} = 2.0276 \pm 0.0003$
		$g_\perp = 2.0114 \pm 0.0003$
$Ca(ClO_3)_2$	O_2^-	$g = 2.0319 \pm 0.0003$
	$ClO-ClO_3^-$	$g_{||} = 2.0277 \pm 0.0003$
		$g_\perp = 2.0115 \pm 0.0003$
$Sr(ClO_3)_2$	O_2^-	$g = 2.0318 \pm 0.0003$
	$ClO-ClO_3^-$	$g_{||} = 2.0276 \pm 0.0003$
		$g_\perp = 2.0114 \pm 0.0003$
$Ba(ClO_3)_2$	O_2^-	$g = 2.0318 \pm 0.0003$
	$ClO-ClO_3^-$	$g_{||} = 2.0276 \pm 0.0003$
		$g_\perp = 2.0114 \pm 0.0003$
$NaClO_3$	O_2^-	$g = 2.0338 \pm 0.0003$
	$ClO-ClO_3^-$	$g_{||}g_\perp = 2.0124 \pm 0.0003$
$KClO_3$	O^-	$g = 2.0338 \pm 0.0003$
	$ClO-ClO_3^-$	$g_{||}g_\perp = 2.0124 \pm 0.0003$

estimated to be 10^{17} cm^{-3}. The intense brown color associated with the radiation induced decomposition was believed to be due to the V center $(ClO-ClO_3^-)$. The g factors are given in Table 4-12.

Hovi and Rasanen[39] examined the IR and UV absorption spectra of irradiated $KClO_3$. Annealing the irradiated crystals at 150°C for 9 hours causes the absorption in the UV region to disappear almost completely. The species identified and their absorption bands are summarized in Table 4-13. Optical bleaching with an intense light source containing the

TABLE 4-13   Infrared bands of $KClO_3$ irradiated and measured at room temperature

Spectral Position (cm^{-1}) 100–4500	Assignment	Radical O_2 and overtones of various radicals
1640–1605	$2\nu_1, \nu_{1+3}$	ClO_2^-
840	ν_3	ClO_2^-
(830)	(ν_3)	$(Cl^{37}O_2^-)$
810	$^2\nu_2$	ClO_2^-
800	ν_1	$Cl^{35}O_2^-$
796	ν_1	$Cl^{37}O_2^-$
406	ν_2	$Cl^{35}O_2^-$
(399)	(ν_2)	$(Cl^{37}O_2^-)$
682	ν	$Cl^{25}O^-$
677	ν	$Cl^{37}O^-$
1030–1270	ν_3	ClO_2
	(ν_1, ν_3)	(O_3)
	ν_3	ClO_4^-
628	ν_4	$Cl^{35}O_4^-$
618	ν_4	$Cl^{37}O_4^-$

wavelength region 380–900 mμ caused a complete disappearence of a radiation produced band at 450 mμ and a slight decrease in a band at 310 mμ. No changes in IR region were observed as a result of optical bleaching. The specie responsible for giving rise to the 450 mμ band was believed to be a radical.

The bands attributed to ClO_4 were observed to increase as a result of thermal bleaching whereas those attributed to ClO_2^- were observed to decrease. The 682 cm^{-1} band was decreased slightly by optical bleaching with 317 mμ light.

Ramasastry and Sastry[140,141,142,143] have made a series of studies on the optical and paramagnetic resonance spectra in irradiated $NaClO_3$ and

$KClO_3$. Irradiation with X-rays produced absorption peaks at 228, 260, 310, 360 and 420 mμ in $NaClO_3$ and at 216, 260, 310, 360, and 450 mμ in $KClO_3$. The 260, 310 and 360 mμ bands are attributed to ClO_2^-, ClO^- and ClO_2 respectively. Paramagnetic resonance spectra indicated the presence of O_3^- and ClO_2. The 420 mμ band in $NaClO_3$ was blieved to consist of two bands superimposed in the region attributed to Cl_2O_6 and O_3^-. The O_3^- band is believed to have its maxima at 420 mμ while that of Cl_2O_6 at 450 mμ Bleaching with light from a tungsten lamp (green filter) caused a decrease in absorption in the region 370–450 mμ with a corresponding increase in the short wave length band at 310 mμ. The authors suggest that the following reactions appear to occur:

$$O_3^- \xrightarrow{h\nu} O_3^-* \qquad\qquad (4\text{-}21)$$

$$Cl_2O_6 \xrightarrow{h\nu} 2\,ClO_3 \qquad\qquad (4\text{-}22)$$

$$2\,ClO_3 + 2\,O_3^-* \longrightarrow 2\,ClO^- + 5\,O_2. \qquad (4\text{-}23)$$

The additional oxygen so produced was qualitatively confirmed by determining the increase in paramagnetic force by the Guoy balance method. Other

TABLE 4-14 A comparison between $NaClO_3$ and $KClO_3$ crystals as regards their optical absorption maxima and peak values in the crystal relative to that of ClO_2^-

| Species | NaClO₃ | | Peak value | KClO₃ | | Peak value |
| | Absorption maximum | | | Absorption maximum | | |
	mμ	eV		mμ	eV	
β-band	228	5.45	—	218	5.69	—
ClO	—	—	—	present		—
ClO_2^-	260	4.75	1	260	4.75	1
ClO^-	322	3.85	0.36	308	4.05	0.59
ClO_2	355	3.5	0.37	360	3.45	0.56
O_3^-	435	2.85	0.37	450	2.75	0.87
Cl_2O_6?	—	—	—	—	—	—

optical bleaching experiments indicated that the ClO_2 radical could be decomposed to $ClO + O$. The resulting ClO radical reacting with the ozonide ion to give ClO_2^- and O_2. In a later study[142] on the effects of optical bleaching of irradiated $KClO_3$ the authors observed that the band attributed to O_3^- ion in irradiated $NaClO_3$ had its peak at 450 mμ. in $KClO_3$ and the existence of Cl_2O_6 (at least in irradiated $KClO_3$) was considered doubtful. In both $NaClO_3$ and $KClO_3$ they observed a strong absorption band in the short wave length region (218–228 mμ) which they assigned to the ClO^- perturbed by the presence of defects in the vicinity. Table 4-14 summarizes the observed absorption maxima in irradiated $NaClO_3$ and $KClO_3$.

The majority of studies on paramagnetic centers formed in irradiated chlorates have been confined to studies at ambient temperature. Patrick and Sargent[144] examined the ESR spectra of $NaClO_3$ irradiated a $-196°C$. All observations were made at this temperature. The spectra were very complex, however, under conditions of high sensitivity and high recorder gain the spectrum with H_0 parallel to the crystal 100 direction consisted of four peaks on scale. The behavior of the spectrum with crystal orientation plus the fact that the isotropic or average g value (2.0076) was in agreement with that reported for the ClO_3 radical found in the perchlorates led the authors to postulate that this center was indeed due to the ClO_3 radical. When the irradiated crystals were warmed to room temperature, the radiation induced blue color changed to amber and the lines attributed to ClO_3 disappeared. The authors were quite convinced that the signal attributed to ClO_3 by Gamble[137] must be some other specie.

4.2.2   Summary of chlorates radiolysis

4.2.2.1   Primary species   The primary species which have been observed and for which there is reasonable agreement are O_3^-, ClO_2, ClO_3 and Cl^-. Room temperature spectra of the irradiated solid reveal the presence of ClO_4^-, ClO^- and ClO_2^-, all of which have been confirmed by chemical analysis. ClO_4^- appears to be a secondary product (see discussion under reference 39), however, there is no evidence that ClO^- and ClO_2^- are formed initially or on the other hand that they arise from reactions of other products (for example $ClO_2 + e \rightarrow ClO_2^-$ or $ClO^- + O$). It can only be stated that at room temperature they are present in crystals irradiated to moderate doses.

4.2.2.2   Kinetics and mechanism of the decomposition   No thorough study on the kinetics of the decomposition of the chlorates has been done. The yield of the products, however, appear to be dose dependent indicating either (or both) radiation or thermally induced reactions of the products. The few annealing studies only reveal the fact that several complex reactions occur (see especially the discussion under reference 128). Both ClO^- and ClO_2^- appear to undergo radiation induced reactions at room temperature (similar to BrO and BrO_2^-), however, no studies have been made as the nature of these reactions. There are good indications that perchlorate ion formation is in some way related to ClO_2^- disappearence. Summarized below are reported G values for the products in the various salts that have been studied.

Summary of G values for the products of the decomposition of the various chlorates

Salt	$G(ClO_4^-)$	$G(Cl^-)$	$G(ClO^-)$	$G(ClO_2^-)$	$G(ClO_2)$	$G(O_2)$	Ref.
$KClO_3$	1.95 (max)	1.3	0.5	2.0	0.2	2.5	127
	1.13	1.64	0.27	2.07	0.21	2.76	130
	1.13	1.64	0.27	2.07	—	—	132
$NaClO_3$	0.22	0.95	0.21	1.34	0.23	1.95	128
	0.22	0.91	0.18	1.32	0.13	2.36	130
$Ba(ClO_3)_2$	0.60	1.72	0.26	2.05	0.25	3.90	130

4.3 Perchlorates

The information available on kinetics of the decomposition of the perchlorates is limited to a very few studies. Heal[57] investigated the decomposition of $KClO_4$ using 50 kVp X-rays as the radiation source. Absorbed dose was determined calorimetrically. Heal only determined the ClO_3^- and Cl^- yields; other products (ClO^- and or ClO_2^-) which were only present in small amounts were expressed as ClO_2^- (oxidizing equivalents). It was observed that Cl^- and ClO_3^- were formed directly from $KClO_4$ in the ratio of 1 mole to 4 moles and not via some intermediate product. The ClO_3^- yield as a function of absorbed dose became constant with possibly some falling off at relatively high absorbed doses, however, the initial yield appeared to be linear. The rate of Cl^- formation showed an abrupt increase after about five hours of radiation and appeared to be constant thereafter; the yield then being approximately linear with absorbed dose. Heal suggested that some

ClO_3^- decomposes to give Cl^-, however, at the absorbed dose where the yield of ClO_3^- levels off to a steady state the rate of formation of Cl^- is constant, indicating that decomposition of $KClO_4$ to form ClO_3^- has reached a steady state or reactions of ClO_3^- other than that to produce Cl^- ion are occuring. The G values for perchlorate disappearance was estimated to be about five.

Baberkin[126] examined the radiation induced decomposition of $KClO_4$ and reported that chlorate and oxygen were the only products formed in un-heated $KClO_4$. The G value for ClO_3^- was reported as 1.1. Heating to 390°C reduced the G value to 0.7. The difference being accounted for by the formation of Cl^- and oxygen.

Hyde and Freeman[145] examined the paramagnetic resonance of am-monium perchlorate irradiated at room temperature with 50 kVp X-rays. Only one center was observed and this was attributed to the NH_3^+ ion. No decay of the resonance spectrum was observed on heating the irradiated sample to 125°C.

Cole[146] reported on paramagnetic centers on single crystals of $KClO_4$ irradiated with X-rays at room temperature. A long-lived stable center was observed and by comparison with the results of Bennett, Ingram and Schonland[147] on ClO_2 in rigid solution he attributed it to the ClO_2 mole-cule.

In a later study, Cole[148] analyzed the paramagnetic center produced in NH_4ClO_4 irradiated at room temperature with 50 kVp X-rays. The spectrum consisted of two sets of lines. One set could be decomposed into an equally spaced triplet of quartets, having relative intensities of $1 : 3 : 3 : 1$. This spectrum arises from the interaction of an unpaired electron with one nucleous of spin 1 and three equivalent nuclei with spin $\frac{1}{2}$. This center was assigned to the NH_3^+ ion. It was confirmed by examination of the completely deuterated molecule which had been identically irradiated. The other center consisted of four evenly spaced lines of equal intensity. From the nature of the hyperfine splitting Cole attributed this center to ClO_3 radical. Radical concentrations were estimated to be about 10^{-3} mole percent.

Atkins *et al.*[134] irradiated a number of perchlorates with cobalt-60 gamma rays. Irradiations were done at room temperature, however, spectra were examined at room temperature and at 77°K. The spectrum of ClO which was well known from studies on photolyzed rigid solutions of ClO_2 was not detected in any of the perchlorates examined (Li, Na, K, Mg, NH_4). The spectrum of ClO_2 was observed in irradiated $KClO_3$ in agreement with

Cole. ClO_3 was detected in irradiated potassium and magnesium perchlorates. In addition to ClO_2 and ClO_3, the spectrum of O_3^- was observed and confirmed by diffuse reflectance spectra (460 mμ). The spectra for various other radicals were detected but could not be resolved.

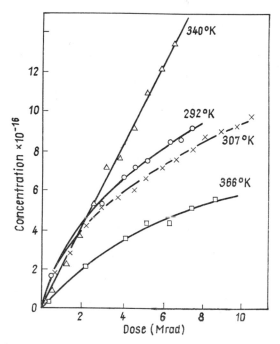

FIGURE 4-7 The relation between the concentration of paramagnetic particles, the dosage, and the temperature of irradiation. (The concentrations are to be reduced by a factor of 1 : 10 for the curve at $T = 366$ K)

Dubovitskii and Manelis[149] examined the kinetics of paramagnetic center formation and disappearence in irradiated $KClO_4$ as a function of temperature. The major effort was concentrated in the temperature range 290–370°K. The radicals detected were ClO_2, ClO_3, O_3^- and an unidentified radical. The total yield of radicals as a function of temperature is shown in Figure 4-7

TABLE 4-15

Temp°K	145°	292°	307°	340°	350°	366°
$G(R)$	0.3	0.04	0.04	0.03	0.1	0.16

8*

and in Table 4-15. The scanning scale for the fixed magnetic field was developed from the hyperfine splitting of DPhPG in benzene solution.

Above 300°K only ClO_2 and O_3^- were observed to exist (plus a small amount of an unidentified center). Above 340°K only O_3^- was significant. The high radical yields observed above 350° were believed due to recombination of less stable radicals and formation of thermally stable radicals. Reaction constants for the disappearance of ClO_2 and O_3^- were determined and are given as:

$$k_{(ClO_2)} = 10^{-16} \exp(-800/RT) \ cm^3/sec \qquad (4\text{-}24)$$

$$k_{(O_3^-)} = 10^{-13} \exp(-1600/RT) \ cm^3/sec. \qquad (4\text{-}25)$$

Fujimoto and Morton[150] examined the EPR spectrum of NH_3^+ formed in irradiated NH_4ClO_4 at 300, 77 and 4°K. The spectra support the idea that the radical is planar and is rotating about an axis perpendicular to the proton plane.

Odian, Acker and Pleitzeke[151] studied the room temperature radiolysis of NH_4ClO_4 using cobalt-60 gamma rays as the radiation source. The major products were observed to be ClO_3^-, Cl_2 ClO^-, no ClO_2 or ClO_2^- was observed. The yields of all products except ClO_3^- were linear with dose. However, stoichiometry was not obtained. The G value, however, for the disappearance of ClO_4^- was far in excess of that which could be accounted for in the observed products, hence, other products were formed in substantial amounts but were not identified. No evidence for NO_2^-, NO_3^- or oxides of nitrogen were obtained. Products which are known to be formed in the thermal decomposition are NH_3, H_2O, and N_2; no observation on these compounds was reported.

The most complete study of the metal perchlorates was that reported by Prince and Johnson[152] on the gamma ray induced decomposition of the alkali and alkaline earth perchlorates. The irradiation source was cobalt-60 and dose was determined using Fricke dosimetry with $G(F_3^{3+}) = 15.46$. No corrections were made for differences in mass absorption coefficients between the dosimeter solution and the salts. All the salts became colored upon irradiation and had a chlorine like odor which was due to ClO_2. The intensity of the color was proportional to the amount of ClO_2 present. Solution of the salts when examined by UV indicated the presence of ClO_2^-, ClO_2 and ClO; X-ray and IR spectra indicated the presence of Cl^- and ClO_3^- in the lattice.

The products observed for the room temperature decomposition and initial G values are summarized in Table 4-16. All of the products appeared to be dose dependent (see Figure 4-8). It was observed that some of the products undergo thermal decomposition even at room temperature; the rate of decomposition of the different products depends on the particular salt and has a complicated temperature dependence. The initial G values shown in Table 4-16 have been corrected for thermal decomposition, but not

TABLE 4-16 Initial G values of alkali and Alkaline earth perchlorates irradiated at room temperature and pressure

Salt	$G_{ClO_4^-}$	$G_{ClO_3^-}$	$G_{ClO_2^-}$	G_{ClO_2}	G_{ClO^-}	G_{Cl^-}	G_{O_2}
Li	3.76	2.80	0.15	0.59	0.10	0.12	2.15
Na	4.36	3.57	0.17	0.11	0.09	0.42	2.96
K	3.83	2.99	0.18	0.12	0.09	0.45	2.68
Rb	5.27	4.06	0.20	0.12	0.14	0.75	3.84
Cs	6.84	5.28	0.22	0.10	0.17	1.07	5.28
Mg	4.67	4.29	0.14	0.07	0.03	0.15	2.62
Ca	4.15	3.44	0.00	0.51	0.08	0.12	1.99
Sr	4.53	3.90	0.19	0.14	0.11	0.19	2.61
Ba	3.20	1.76	0.84	0.42	0.12	0.06	2.18

that which occurred during irradiation. Careful tests were made for the presence of ozone and free chlorine but none was observed. Stoichiometry calculations made on the basis of $G^0(O_2) = \frac{1}{2}G^0(ClO_3^-) + G^0(ClO_2) + G^0(ClO_2^-) + 3/2\, G^0(ClO^-) + 2\, G^0(Cl^-)$ gave a good stoichiometric relationship between experimental and calculated $G(O_2)$ values for all the

TABLE 4-17 A comparison of experimental and calculated G^0O_2 values

Compound	G^0O_2, exptl.	G^0O_2, calcd.	Variation, %	
			before correction	after correction
LiClO$_4$	2.15	2.24	+18	+4
NaClO$_4$	2.96	2.98	+3	+1
KClO$_4$	2.68	2.78	+6	+4
RbClO$_4$	3.84	3.99	+6	+4
CsClO$_4$	5.28	5.28	+1	0
Mg(ClO$_4$)$_2$	—	2.64	—	—
Ca(ClO$_4$)$_2$	1.99	2.33	+30	+17
Sr(ClO$_4$)$_2$	2.61	2.74	+8	+5
Ba(ClO$_4$)$_2$	2.18	2.23	+12	+2

salts except Li, Ba and Ca perchlorates. $Mg(ClO_4)_2$ was not examined because of the experimental difficulties due to the hygroscopic nature of the salt. It was observed that the poorest stoichiometry was found with those salts that had the highest $G(ClO_2)$ values. This led the authors to assume that associated with the formation of ClO_2 was a metal oxide or peroxide. When a correction for oxygen made on this assumption was applied to the stoichiometry calculations excellent agreement between experimental and calculated values was obtained for all salts except $Ca(ClO_4)_2$. The discrepancy with this salt was attributed to experimental difficulties. The results are summarized in Table 4-17. An unidentified product was believed to be present which, had the properties of reacting with KI at pH9 and was either volatile or capable of being destroyed by flash boiling. The specie was tentatively assigned to ClO_3. G values for the specie was 0.1 in $KClO_4$, $RbClO_4$, and $CsClO_4$; 0.2 in $LiClO_4$ and 0.0 in $NaClO_4$. A typical plot of product yields versus absorbed dose is shown in Figure 4-8.

The decomposition as a function of irradiation temperature was examined only for $KClO_4$. These results are summarized in Table 4-18. As is apparent, ClO^- is not a primary product, and it appears to be associated in some way with ClO_2 i.e. the decrease in $G(ClO_2)$ is accompanied by an increase in $G(ClO^-)$. Annealing studies on irradiated perchlorates

TABLE 4-18 G^0 yields of $KClO_4$ irradiated at various temperatures and atmospheric pressure

Temp., °C.	$G^0_{ClO_4^-}$	$G^0_{ClO_3^-}$	$G^0_{ClO_2^-}$	$G^0_{ClO_2}$	$G^0_{ClO^-}$	$G^0_{Cl^-}$	$G^0_{O_2}$
−196	4.0	2.7	0.58	0.18	0.00	0.5	3.0
−80	4.0	2.8	0.49	0.17	0.00	0.6	2.7
−16	3.7	2.8	0.26	0.06	0.06	0.5	2.4
−8	3.5	2.6	0.26	0.09	0.06	0.5	2.6
0	3.4	2.6	0.20	0.70	0.06	0.5	2.2
19	3.7	2.9	0.20	0.10	0.06	0.5	
20	3.7	2.9	0.19	0.09	0.08	0.5	2.5
72	4.1	3.3	0.19	0.09	0.00	0.5	2.7
260	4.9	4.2	0.00	0.00	0.00	0.7	2.0
295	3.8	2.9	0.00	0.00	0.00	0.9	1.0

revealed a very complicated concentration dependence for the species involved. It was observed that ClO^- was generally more stable thermally in the alkaline earth lattices than in the alkali metal lattices and that ClO_2 and ClO_2^- are least stable in the cesium and rubidium lattices. Analysis of

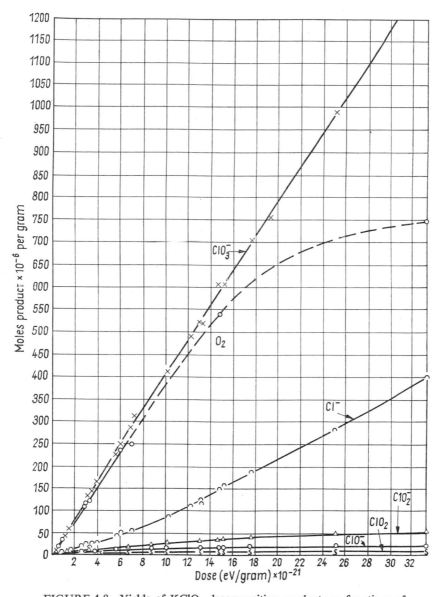

FIGURE 4-8 Yields of KClO$_4$ decomposition products as functions of dose. O = O$_2$, × = ClO$_3^-$, ▲ = ClO$_2^-$, O = ClO$^-$, s = ClO$_2$, ∩ = Cl$^-$

irradiated $KClO_4$ annealed at different temperatures is shown in Table 4-19. It was also observed that solutions of most of the irradiated perchlorates had an alkaline pH which was in excess of that which could be attributed to hydrolysis of the ClO_2^- and ClO^- present. This alkaline pH was attributed to the metal oxide or peroxide (which had been presumed to accompany ClO_2 formation) and was eliminated by annealing the irradiated salts at 170°. Chlorine balance studies indicated that a neutralization reaction had occurred probably via a reaction of ClO_2 and the metal oxide or peroxide.

Kinetic analysis of the data revealed that back reaction of ClO_3^- to form ClO_4^- was occurring in all the salts examined. It was also observed that in some salts ClO_2 was converted on standing at room temperature to ClO_2^-, indicating diffusion of some product in the lattice. The overall mechanism for the decomposition was given as:

$$ClO_4^- \rightsquigarrow ClO_3^- + O \qquad (4\text{-}26)$$

$$ClO_4^- \rightsquigarrow Cl^- + 2\,O_2 \qquad (4\text{-}27)$$

$$ClO_4^- \rightsquigarrow [ClO_4^-]^* \qquad (4\text{-}28)$$

$$[ClO_4^-]^* \longrightarrow Cl^- + 2\,O_2 \qquad (4\text{-}29)$$

$$\longrightarrow ClO_2^- + O_2 \qquad (4\text{-}30)$$

$$\longrightarrow ClO_2 + O^- + O \qquad (4\text{-}31)$$

$$[ClO_4^-]^* + \text{thermal activation} \longrightarrow ClO^- + O_2 + O \qquad (4\text{-}32)$$

$$ClO_4^- + O \longrightarrow ClO_3^- + O_2 \qquad (4\text{-}33)$$

$$ClO_3^- + \tfrac{1}{2}O_2 \rightsquigarrow ClO_4^-. \qquad (4\text{-}34)$$

The mechanism of the decomposition was developed under the following considerations:

1) ClO_3^- is a major product in the decomposition hence reactions (4-26) and (4-33) seem reasonable.

2)   Initial Cl^- yields appear to be independent of temperature indicating that thermal activation is not necessary for its formation hence reaction (4-27).

3)   The appearance of ClO^- with a simultaneous decrease in ClO_2^- and ClO_2 suggests a common precursor hence reactions (4-29), (4-30) and (4-31).

4)   There also appears to be a relation between $G(Cl^-)$ and $G(ClO_2^- + ClO_2)$. High $G(Cl^-)$ values are accompanied by low $G(ClO_2^- + ClO_2)$

TABLE 4-19 Analysis of irradiated $KClO_4$ heated at different temperatures

Dose, eV/g × 10^{-21}	Temp., °C	Heating time min.	O_2, μmoles/g.	Cl^- μmoles/g.	ClO^- μmoles/g.	ClO_2^- μmoles/g.	ClO_2 μmoles/g.	ClO_3^- μmoles/g.	ClO_4^- μmoles/g.	Wt. loss %
33.5	R.t.	0	562 ± 6	409 ± 6	18.2 ± 1.0	52.7 ± 1.0	8.8 ± 0.2	1269 ± 13	5444 ± 20	
	51	5498	552 ± 6	442 ± 5	2.7 ± 0.1	31.2 ± 0.6	0.8 ± 0.1	1273 ± 12	5648 ± 20	0.16
	84	2401	369 ± 4	464 ± 5	0.0	9.5 ± 0.2	0.5 ± 0.1	1267 ± 13	5647 ± 20	0.94
	151	210	244 ± 3	469 ± 5	0.0	0.3	0.0	1278 ± 13	5650 ± 20	1.43
14.7	R.t.	0	543 ± 6	149 ± 2	11.8 ± 0.3	33.7 ± 0.7	5.3 ± 0.1	611 ± 6	6410 ± 10	
	51	5498	555 ± 6	162 ± 2	5.3 ± 0.1	26.5 ± 0.5	1.2 ± 0.1	604 ± 6	6422 ± 10	0.03
5.90	R.t.	0	238 ± 3	48.8 ± 1.0	6.2 ± 0.1	14.3 ± 0.3	4.8 ± 0.1	245 ± 3	6898 ± 5	
	84	2401	238 ± 3	54.4 ± 1.0	3.0 ± 0.1	12.8 ± 0.3	1.6 ± 0.1	245 ± 3	6901 ± 5	0.00
	151	210	233 ± 3	60.5 ± 1.2	0.3	7.3 ± 0.2	0.8	247 ± 3	6902 ± 5	0.02
6.87	R.t.	0		54.8 ± 0.8	8.5 ± 0.1	17.7 ± 0.4	4.9 ± 0.1			
	52	1110		57.5 ± 0.8	6.8 ± 0.1	17.7 ± 0.4	4.1 ± 0.1			

values. These considerations also favor a common primary specie such as ClO_4^-* (or ClO_4^{2-}).

Oblivantsev, Lykhin and Boldyrev[153] who reported on the gamma ray decomposition of potassium, and cesium perchlorates, observed G values for the decomposition comparable to those found by Prince and Johnson.[152]

In a later study, Oblivantsev et al.[154] examined the role of thermal track effects on the decomposition of potassium, rubidium and cesium perchloration using 4.7 MeV protons. Absorbed dose was carefully determined by calculating the stopping powers using the Bethe–Bloch formula for heavy charged particles and using this information to determine the energy absorbed by the sample for known ion currents. The calculated mean absorbed dose rates were $0.5–1.5 \times 10^{19}$ eV/g. G values for perchlorate decomposition were 6.0, 6.5 and 7.4 for $KClO_4$, $RbClO_4$ and $CsClO_4$ respectively. Chloride yields in the same order were 1.2, 1.7 and 1.9. Chlorite and hypochlorite were analyzed for and found to be comparable with those obtained for gamma irradiated samples.[152,153] The authors argue that since it has been established that both hypochlorite and chlorite are thermally unstable (see Table 4-19) it would appear that thermal track effects are not operable in the proton radiolysis of these perchlorates i.e. the difference in G values for proton radiolysis of these perchlorates as compared to gamma radiolysis cannot be explained on the basis of increased temperature due to the high LET.

Belevskii et al.[155] have observed the presence of ClO_3 and $HClO_4$ at $-195°C$ in irradiated anhydrous perchloric acid. The assignment of the signal attributed to $HClO_4^+$ was determined largely by the fact that this signal disappeared at about $-140°C$ and a signal definitely assigned to ClO_2 appeared. The authors postulated that $HClO_4^+$ ion decomposed at this temperature to give ClO_2. The signal attributed to ClO_3 was observed to disappear at about $-120°C$. As the temperature was raised the irradiated crystals acquire an intense color which is believed to be due to the formation of Cl_2O_6.

4.3.1 Summary of perchlorate radiolysis

Although the available data are limited it does appear that sufficient information has been obtained to form some good conclusions at least on the alkali and alkaline earth perchlorates.

4.3.1.1 Primary species
ESR and optical spectra definitely indicate that ClO_2, O_3^- and ClO_3 are present at ambient temperature. These results have been confirmed by three different laboratories (reference 146, 147, 135,

149). Chemical and optical studies definitely support the presence of ClO_2, Cl^-, ClO_3^- and ClO_2^-. It must be added that no optical studies at low temperatures have been done hence it is quite possible that ClO_2^- is not a primary product. We only know that room temperature analysis of samples irradiated at $-195°C$ show the presence of ClO_2^-. In addition, evidence supporting the presence of ClO_3 (or possibly O_3^-) was also obtained.[152] No evidence for ClO or ClO^- as a primary product has been obtained.

It appears therefore that Cl^-, ClO_3^-, ClO_2, ClO_2^-, (?) O_3^- and ClO_3 are primary products in the radiation induced decomposition of the alkali and alkaline earth perchlorates. In NH_4ClO_4, NH_3^+ and ClO_3^- were the only primary species positively identified. No ClO_2^- or ClO_2 was observed to be present.

4.3.1.2 Kinetics of the decomposition The decomposition is not linear with absorbed dose i.e. $G(ClO_3^-)$ (as well as the other products) is dose dependent. The products of the decomposition undergo complex thermal and radiation induced reactions including a back reaction of ClO_3^- to form ClO_4^-. There does not appear to be any effect of intensity in these compounds. There appears to be a substantial increase in $G(-ClO_4^-)$ with increasing LET however, there is no effect on product distribution.

4.4 Iodates and periodates

The available literature on the radiation induced decomposition of these salts is very limited, the major effort being confined to the chemistry of recoils following neutron capture. Only these studies bearing on the radiation induced decomposition will be discussed.

Cleary, Hamill and Williams[156] in their studies on the chemical consequences of neutron capture by iodine in Li, Na, K, and NH_4^+ iodates observed that the majority of the[128] I atoms formed were in the reduced state (as I_2 or I^-). The retention as iodate increased if the neutron bombarded samples were exposed to ionizing radiation indicating a radiation induced reaction between iodide (or iodine) and oxygen to form iodate.

Walton and Croal[156] studied the decomposition of potassium iodate using fission recoils. Uranium oxide and KIO_3 were ground together in a mortar to a uniform powder and exposed as such to mixed pile radiation at a neutron flux of 1.6×10^{12} neutrons/cm² sec. in stoppered silica tubes. Free condensed iodine was observed on the walls of the unopened tubes.

Analysis of aqueous solution of the irradiated salt was confined to the determination of the total reduced formed of iodine and was reported as I^- (i.e. no analysis for IO^- or IO_2^- was made) and iodine. G value for the decomposition of iodate was estimated to have a minimum value of 2 and a maximum value of 6.

Mohanty[157] irradiated crystalline KIO_3 with 2 MeV electrons. The only product detected was I^- with a very low yield. He concluded that KIO_3 is very stable toward the action of ionizing radiation.

The most recent report on recoil studies of KIO_3 and KIO_4 summarizes the chemistry of these systems and does include some data on the radiation induced decomposition.[158] KIO_3 appears to be exceptionally stable toward decomposition by cobalt-60 gamma rays. After exposure to neutron irradiation radioiodide and radioiodate are the only species observed, no radio hypoidite (IO^-) or radioiodite (IO_2^-) were observed, however, a small amount of radioperiodiate (IO_4^-) was observed. Annealing studies a neutron irradiated KIO_3 show that iodide ion is easily oxidized to iodate and this reaction occurs at temperatures as low as $-80°C$. Thermal annealing also produced IO_4^- from IO_3^- at temperatures above $150°C$. It was also observed that irradiation induced annealing occurred i.e. I^- was converted to IO_3^- when the neutron irradiated salt was exposed to cobalt-60 gamma rays.

The radioactive products after exposure of KIO_4 to neutron irradiation are radioidide, radioiodate and radioperiodate. KIO_4 appears to decompose with ease to give I^-, IO_3^-, and O_2. Thermal annealing studies on irradiated KIO_4 showed that I^- was rapidly converted to IO_3^- and IO_3^- is converted to IO_4^-. Exposure of the neutron irradiated salt to cobalt-60 produced a radiation annealing reaction similar to the thermal annealing process.

The fact that I^- is readily converted to IO_3^- in neutron irradiated KIO_3 readily explains the stability of this salt toward radiation damage and the rapid oxidation of IO_3^-, from KIO_4 decomposition, to IO_4^- explains the formation of IO_4^-.

It appears from the available data that further studies on these salts should be most revealing, especially those at low temperatures.

References

105   G. E. Boyd, J. W. Cobble and S. Wexler, *J.A.C.S.* **74**, 237 (1952)

106   J. W. Cobble and G. E. Boyd, *ibid.* **74**, 1282 (1952)

107   G. E. Boyd and J. W. Cobble, *J. Phys. Chem.* **63**, 919 (1959)

108 A. G. Maddock and H. Muller, *Trans. Far. Soc.* **56**, 509 (1960)

109 G. E. Boyd, E. W. Graham and Q. V. Larson, *J. Phys. Chem.* **66**, 300 (1962)

110 T. Andersen, *Nature* **200**, 4911 (1963)

111 G. E. Boyd and Q. V. Larson, *J. Phys. Chem.* **68**, 2627 (1964)

112 G. E. Boyd and T. G. Ward, Jr., *J. Phys. Chem.* **68**, 3809 (1964)

113 C. Ramasastry and Y. V. G. S. Murti, *Ind. J. of Pure and Appl. Physics* **2**, 35 (1964)

114 G. E. Boyd and G. V. Larson, *J. Phys. Chem.* **69**, 1413 (1965)

115 J. W. Chase and G. E. Boyd, *J. Phys. Chem.* **70**, 1031 (1966)

116 J. W. Chase and G. E. Boyd, "Symposium on chemical and physical effects of high energy radiation on inorganic substance" *ASTM STP* **400**, Am. Soc. Testing Mat., 1966

117 T. Andersen, H. E. Lundager Madsen and K. Olesen, *Trans. Far. Soc.* **62**, 2409 (1966)

118 P. J. Herley and P. W. Levy, *J. of Chem. Phys.* **46**, 627 (1967)

119 T. Andersen, J. R. Byberg and K. J. Olsen, *J. Phys. Chem.* **71**, 4129 (1967)

120 F. T. Gamble, *J. Chem. Phys.* **47**, 1193 (1967)

121 G. E. Boyd and Q. V. Larsen, *J.A.C.S.* **90**, 254 (1968)

122 L. C. Brown, G. M. Begum, and G. E. Boyd, *J.A.C.S.* **91**, 2250 (1969)

123 E. V. Apelman, *J.A.C.S.* **90**, 1900 (1968)

124 K. J. McCallum and O. G. Holmes, *Can. J. Chem.* **29**, 691 (1951)

125 L. J. Sharman and K. J. McCallum, *J. Chem. Phys.* **23**, 597 (1955)

126 A. S. Baberkin, "Trudy Pervago Vsesoyuz Soveschchaniya Po Radiatsion Khim", *Akad Nauk S.S.S.R. Otdel, Khim Nauk*, Moscow, 167–8 (1957) publ. 1958

127 H. G. Heal, *Can. J. Chem.* **37**, 979 (1959)

128 P. F. Patrick and K. J. McCallum, *Nature* **194**, No. 4830 (1962)

129 C. E. Burchill M. A., *Thesis Univ. of Saskatchewan* (1960)

130 C. E. Burchill, P. F. Patrick and K. J. McCallum, *J. Phys. Chem.* **71**, 4560 (1967)

131 G. E. Boyd and Q. V. Larson, *J.A.C.S.* **90**; 19, 5092, 1968

132 L. E. Brown and G. E. Boyd, *J. Phys. Chem.* **73**, 396 (1969)

133 T. E. Hasty, W. B. Ard, and W. G. Moulton, *Phys. Rev.* **116**, 1459 (1959)

134 P. W. Atkins, J. A. Brivati, N. Keen, M. C. R. Symons and P. A. Trevalion, *J. Chem. Soc.* **4785** (1962)

135   W. McLachlan, M. C. R. Symons and P. Townsend, *J. Chem. Soc.* **952** (1959)

136   R. W. Holmberg, ORNL-3320 (1962) TID-4500 p. 106–107

137   F. T. Gamble, *J. Chem. Phys.* **42**, 3542 (1965)

138   T. Cole, *J. Chem. Phys.* **35**, 1169 (1961)

139   G. I. Subbotin, *Optics and Spectroscopy Engl. Transl.* **18**, 92 (1965)

140   C. Ramasastry and S. B. S. Sastry, *J. Phys. Soc. Japan* **18**, 1220 (1963)

141   C. Ramasastry and S. B. S. Sastry, *Ind. J. Pure and Appl. Phys.* **3**, 414 (1965)

142   C. Ramasastry and S. B. S. Sastry, *J. Phys. Chem. Solids* **29**, 399 (1968)

143   C. Ramasastry, S. B. S. Sastry, Y. V. G. S. Murti and J. Sobhandri, *J. Phys. Soc. Japan* **19**, 770 (1964)

144   P. F. Patrick and F. P. Sargent, *Can. J. Chem.* **46**, 1818 (1968)

145   J. S. Hyde and E. S. Freeman, *J. Phys. Chem.* **65**, 1636 (1961)

146   T. Cole, *Proc. Nat'l Acad. Sci. U.S.* **46**, 506 (1960)

147   Bennett, Ingram and Schonland, *Proc. Phys. Soc.* **69**, A 556 (1956)

148   T. Cole, *J. Chem. Phys.* **35**, 1169 (1961)

149   A. V. Dubovitskii and G. B. Manelis, *Kinetika i Kataliz* **6**, 828 (1965) (*Kinetics and Catalysis* **6**, 747 (1965))

150   M. Fujimoto and J. R. Morton, *Can. J. Chem.* **43**, 1012 (1965)

151   G. Odian, T. Acker and T. Pletzke, *J. Phys. Chem.* **69**, 2477 (1965)

152   L. A. Prince and E. R. Johnson, *J. Phys. Chem.* **69**, 354 (1965); ibid 69, 377 (1965)

153   A. N. Oblivantsev, V. M. Lykhin and U. V. Boldyrev, *Journal of All Union Chem. Soc.* (*Zh. VKhO*) *No.* **5**, 598 (1965)

154   A. N. Oblivantsev, V. V. Boldyrev, L. I. Eremin and V. M. Lykhin, *Kinet. and Cataly.* **7**, (6) 1015 (1966)

155   V. N. Belevskii, L. T. Bugaenko and V. Ya. Rosolovskii, *Russ. J. of Phys. Chem.* **42**, 1515 (1968)

156   R. E. Cleary, W. H. Hamill and R. R. Williams, *J.A.C.S.* **74**, 4675 (1956)

157   G. N. Walton and I. F. Croall, *J. Inorg. and Nucl. Chem.* **1**, 149 (1955)

158   S. R. Mohanty, *Ind. J. of Chem.* **2**, 205 (1964)

5

Radiation induced decomposition of the azides, sulfates, carbonates, and permanganates

With the exception of the azides there has been relatively little research reported on the radiation induced decomposition of these classes of compounds; however there has been a fair amount of research on the paramagnetic centers formed during irradiation in these compounds. A general review is given by Symons[159] on these compounds and on others not discussed in this text. In what follows we will discuss the total work reported on each class of compounds and if possible summarize the existing knowledge.

5.1 Azides

A large fraction of the research reported on the azides has been those studies related to the photochemical decomposition. In general these studies postulate the formation of N_3 radicals at cation vacancies with subsequent decomposition of the radicals to give N_2. An interesting and informative early theoretical study by Mott[160] is worth reading. A very early study on the decomposition of barium azide by X-rays was reported by Gunther, Lepin and Andreew.[161] They observed decomposition equivalent to 50–60% when this salt was exposed to 50 kVp X-rays.

Thomas and Tompkins[162] studied the photodecomposition of barium azide irradiated with a low pressure mercury arc (2537 Å) as a function of temperature and intensity. They observed that the rate of decomposition was constant at constant temperature and intensity and varied as the square of the intensity at constant temperature. They also observed that the rate increased in a complex manner with temperature at constant intensity. A plot of log (rate) against $1/T$ produced two linear curves which intersected at a point corresponding to a temperature of $-52°C$. The apparent activation energy below $-52°C$ was 500 cal/mole and above this temperature 4500 cal/ mole. The ultimate decomposition was considered to involve two azide radicals ($2 N_3^- \rightarrow 3 N_2$) since the photodecomposition was proportional to the intensity squared. The initial rate of the decomposition at constant temperature, however, was proportional to the first power of the initial intensity. The initial photolytic rate decreased slowly with exposure and became constant with a rate proportional to the square of the intensity. The observed rate therefore, consists of two parts: azide ions on the surface react with excitons to form azide radicals, the electrons being trapped at impurity centers. The positive holes formed remain on the surface. Reaction occurs between two trapped adjacent radicals and the reaction speads over the surface. As the products spread over the surface the number of places where excitons can be trapped increases. The increase reaches a maximum when the surface is completely covered. When this process has occurred there is a constant area of reaction and the rate is constant.

It was also observed that there appeared to be two types of trapping sites, one that requires zero activation energy for decomposition of excited azide ions and one which did require some thermal activation. It was suggested that the former could be a cation vacancy and the latter some other imperfection. In this latter case the presence of the Ba^{2+} ion will produce a potential barrier which the "active centers" would have to surmount to approach each other for decomposition. A rate expression consistent with these observations was developed which reproduced the experimentally determined rate of decomposition as a function of temperature quite satisfactorily.

$$R = \text{const.} [1.9 \times 10^{-5} + e^{-5350/RT}]. \tag{5-1}$$

In a subsequent paper, Jacobs and Tompkins[163] studied the photodecomposition of potassium azide and confirmed their observation on barium azide. In the studies on KN_3 they observed essentially no photocurrent and therefore believed that the process

$$h\nu + N_3^- \longrightarrow \text{(positive hole)} + \text{electron (in conduction band)} \tag{5-2}$$

did not occur. Their data supports the view that excitons are trapped at cation vacancies as shown in equation (5-3).

$$h\nu + N_3^- \longrightarrow N_3^-{}^* \quad \text{(excited azide ion).} \tag{5-3}$$

The previous studies on barium azide decomposition as a function of temperature had indicated the possibility of two kinds of traps, one in which decomposition of excited azide ions occurred with essentially zero activation energy and the other which requires some thermal activation for decomposition of the excited azides ions. The studies on potassium azides, however, led the authors to postulate that only one type of exciton trap exists (cation vacancies) but due to the presence of steric factors some thermal activation is necessary i.e. excited ions at trapping sites have a low probability of decomposing with zero activation energy, but as the excited ions (which form a complex) obtain thermal energy from the lattice the steric factor decreases. The temperature dependance of the rate of decomposition had the same form as that for the barium salt. A plot of log rate vs $1/T$ gave two linear curves which intersected at a temperature corresponding in this case to 1°C. The apparent activation energy below 1°C was 700 cal/mole and above 1°C 2700 cal/mole.

Heal[57] studied the decomposition of sodium azide, using 50 kVp X-rays as the radiation source, as a function of intensity and temperature. The products of the decomposition were determined by chemical analysis after dissolution of the irradiated salt in water. The products observed were N_2, OH^- and NH_3 in the stoichiometric ratio of 4:3:1. The actual ratio values for N_2, OH^- and NH_3 were 1:0.81:0.22 at 51°C and 1:0.72:0.25 at 102°C. In addition to these products a small amount of hydrazine was also detected. The G values for the decomposition of NaN_3 observed were 5.2 ± 1 at 102°C and 4.0 ± 1 at 52°C. The apparent activation energy for the decomposition was 500 cal/mole. No effect of a 3 fold variation in intensity was observed. A mechanism for the decomposition was postulated wherein the initial act was believed to be dissociation of N_3^- to yield:

$$N_3^- \xrightarrow{\hspace{0.8cm}} N_2 + N^- \tag{5-4}$$

followed by reactions of N^- with the solvent and dissolved azide ion:

$$N^- + H_2O \longrightarrow NH + OH^- \tag{5-5}$$

$$NH + N_3^- \longrightarrow NH^- + 3/2\,N_2 \tag{5-6}$$

$$NH^- + H_2O \longrightarrow NH_2 + OH^- \tag{5-7}$$

$$NH_2 + N_3^- \longrightarrow NH_2^- + 3/2\,N_2 \tag{5-8}$$

$$NH_2^- + H_2O \longrightarrow NH_3 + OH^-. \tag{5-9}$$

This mechanism accounts for the small amount of hydrazine observed and stoichiometric ratio of the products. In addition, since about 25% of all excited azide ions decomposed Heal believed that single exciton mechanism was involved rather than the double exciton mechanism postulated by Tompkins *et al.* in the photodecomposition.

Groocock and Tompkins[164] studied the decomposition of sodium and barium azides using 100–200 kV electrons. There was no darkening of the salts after exposure to the electron beam indicating the absence of metallic nuclei (Heal made a similar observation when NaN_3 was irradiated with X-rays). The initial act was considered to be ionization of azide ions with the electrons ejected into the conduction band or out of the crystal. The holes so formed capturing an electron or reacting with an adjacent excited azide ion, the pair decomposing to give nitrogen; the anion vacancies formed capture an electron and form an *F* centre. It was believed that decomposition occurs primarily on the surface, and that positive holes formed in the bulk of the crystal will largely recombine with electrons to regenerate azide ions. The decomposition at the surface was considered to occur by a bimolecular reaction of adjacent positive holes, however, a small amount of unimolecular decomposition was not excluded. A theoretical equation in good agreement with experiment was developed. In addition to these studies the effect of pre-irradiation on the thermal decomposition was also studied. In all of these studies, however, only gas evolution was determined.

Bowden and Singh[165] studied the effect, of neutron irradiation, fission recoils, X-rays and high energy electrons on the explosive properties of lead, silver and cadmium azide. No chemical analysis or other physical data were obtained. The results showed no effect on the initiation rate of explosion on these substances. A similar result was obtained by Muraour and Ertaud.[166]

Rosenwasser, Dreyfus and Levy[167] examined the radiation induced coloring in sodium azide using reflectance spectrometry. After exposure to gamma-rays crystals of sodium azide turned a brown color whose intensity appeared to be proportional to absorbed dose and also developed three absorption bands (1) 3600 Å (2) 7600 Å and (3) 6600 Å. The 3600 Å band was found to be relatively stable at 20°C. On standing at this temperature the 3600 Å band decreases and the 6600 Å and 7600 Å bands disappear. This is considered evidence that these bands are involved in some electron transfer process. On exposure to fast neutrons a band at 6000 Å develops which is greatly enhanced by heating at 125°C. This band also appears when crystals exposed to gamma rays are heated to 150–240°C. Heating unirradiated crystals produces a band near 3600 Å (it may be recalled that Heal also observed an absorption band at 3400 Å).

Heal[168] in a continuation of his studies on the radiation induced decomposition of sodium azide established (contrary to his previous work) that no nitride ions are produced during the decomposition. This was established by dissolving the irradiated crystals in liquid ammonia or in mercurous chloride solutions. Samples of sodium azide irradiated with X-rays and subsequently dissolved in liquid ammonia at −78°C produced the characteristic blue color of a solution of sodium metal in this solvent plus the evolution of N_2. Although this experiment did not distinguish between trapped electrons and sodium atoms it did provide sound evidence that nitride ions are not produced in quantity during the decomposition because nitride ion would be rapidly converted to the colorless amide ion. Heal believed these results supported the idea that single azide ions do not decompose via:

$$N_3^- \longrightarrow N_2 + N + e^- \qquad (5\text{-}10)$$

since the resulting nitrogen atoms would readily capture electrons to give N^{3-}. The formation of ammonia found in the previous studies when irradiated azide ions were dissolved in water was believed to occur from the reduction of azide ions by H atoms produced when either trapped electrons or sodium atoms react with water

$$Na + H_2O \longrightarrow Na^+ + OH^- + H \qquad (5\text{-}11)$$

$$e^- + H_2O \longrightarrow OH^- + H. \qquad (5\text{-}12)$$

The blue color formed when the irradiated salt was dissolved in liquid ammonia fades when the solution is warmed to −60°C and this fading is attributed to reduction of azide ions by solvated electrons.

The rate of decomposition was found to be proportional to the first power of the intensity over the entire temperature range studied (−186°C to 202°C). A plot of G versus temperature produces a curve which is most easily explained as the resultant of two process (see Figure 5-1). One process, which required no thermal activation, was believed to be the formation of an azide radical which then reacts with an azide ion according to:

$$N_3 + N_3^- \longrightarrow 3 N_2 + e^-. \qquad (5\text{-}13)$$

The other process was described as being due to excitons that are initially trapped at lattice imperfections being thermally activated to electrons in

9*

the conduction band and positive holes. The azide radicals formed, then react according to equation (5-11). Using the value obtained by Tompkins for the mean life of an exciton Heal obtained a theoretical expression which gave G (total) as a function of temperature. This expression is plotted as the solid line in Figure 5-1.

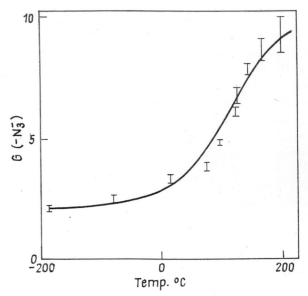

FIGURE 5-1 $G(-N_3^-)$ as a function of temperature. Solid line represents theoretical curve

The decomposition of some alkali azides using cobalt-60 gammas as the radiation source was studied by Boldyrev *et al.*[169] They observed a large descrepancy in the free metal yield relative to the azide ion decomposed and attributed some of this discrepancy to the formation of nitrite ion via a radiation induced reaction of metal with air. This discrepancy was most noticible in rubidium and cesium azides (see Figure 5-2). The authors observed a linear relation between G for azide decomposition and "free" space and propose that "free" volume is a significant factor in the decomposition of these salts. Thermal decomposition temperatures, G values and other pertinent data are summarized in Table 5-1.

Oblivantsev *et al.*[154] studied the decomposition of Na, K, Rb and Cs azides using 4.7 MeV protons as the radiation source with the primary

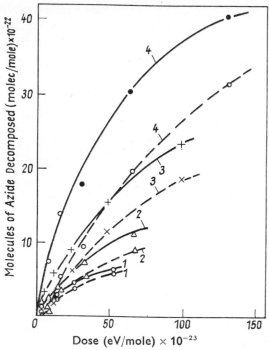

FIGURE 5-2 Radiolysis of alkali metal azides as a function of absorbed dose. Solid lines show the decomposition of azide ion and dotted lines the formation of free metal. (4 = Rb, 3 = Cs, 2 = K, 1 = Na)

purpose of understanding the effect of LET. *G* values for the decomposition of these salts were appreciably less using heavy particle radiations than using gamma rays; the initial yields being 1.7, 2.6, 4.1, and 6.6 for the Na, K, Rb and Cs azides respectively (see Table 5-1 for gamma ray values). As

TABLE 5-1   Correlation of *G* Values with other physical property data

Cation	Cationic radius, Å	Ionization potential, eV	Free volume of elementary crystal cell, A	Decomposition temperature, °C, Å	Initial radiolytic yield, molecules/ 100 eV
Na⁺	0.98	5.138	26.1	240–275	2.4
K⁺	1.38	4.339	28.4	222–255	3.95
Rb⁺	1.48	4.176	34.1	–355	6.05
Cs⁺	1.69	3.893	40.7	–395	10.0

with the decomposition induced by gamma rays a large descrepancy between the yield of the metal and azide decomposition, as required by stoichiometry, was also observed. The authors believe their results on azide radiolysis are not compatable with the idea that differences in yields obtained with high LET sources can be explained by thermal track effects since initial G values were lower for proton irradiation. If thermal track effects were operative then of course one should expect appreciably higher yields for proton irradiation. However it is difficult to understand why G values for the decomposition of the azides should be less with proton irradiation than with gamma irradiation. This would appear to indicate that either the decomposition was extremely complex involving many competitive reactions or the dosimetry is in error. If the reaction is indeed very complex then one might expect some effect of intensity, which has not been observed (see references 57 and 168). We are inclined to think therefore that there is an error in dosimetry and that the G values for proton irradiation are at least equal to those for gamma irradiation. This view is supported by the fact that the ratio of the G values obtained with the two sources is constant, i.e. G (gamma)/G (proton) for all the azides studied is constant.

Zakharov *et al.*[170] observed an accelerating effect on the initial rate of radiolytic decomposition of lead azide by the addition of Ag^+. Single crystals of lead azide were prepared containing varying amounts of silver ion and were exposed to 200 kVp X-rays at a dose rate of 3.2×10^{16} eV/g/sec. It was observed that a large fraction of the evolved nitrogen remained trapped in the lattice, hence nitrogen yields were determined by dissolving the irradiated crystals in dilute nitric acid. A very substantial increase in the rate of decomposition as a result of the addition of silver ion occurs. This is shown in Figure 5-3. Silver ion forms a solid solution in this system and the authors explain the acceleratory effect of this ion by postulating that decomposition occurs by electrons being trapped in anion vacancies and that silver ion increases the number of these vacancies. The larger the number of vacancies the greater is the probability that the trapped electrons will not recombine with holes to form azide ions.

The mechanism postulated is:

$$N_3^- \rightsquigarrow N_3 + e^- \tag{5-14}$$

$$N_3 + e \longrightarrow N_3^- \tag{5-15}$$

$$e^- + X \longrightarrow X^- \tag{5-16}$$

$$2\,N_3 \longrightarrow 3\,N_2 \tag{5-17}$$

where X = anion vacancy.

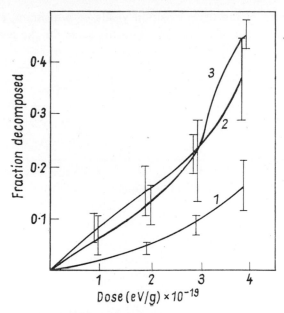

FIGURE 5-3   Dependance of extent of decomposition on absorbed dose for pure PbN_6 (1), PbN_6 containing $1.5 \, mol\%$ Ag (2), PbN_6 containing $5 \, mol\%$ Ag (3)

5.1.1   Summary of azide decomposition

Although the available literature on the decomposition of the azides is meagre some interesting data have been obtained. It does appear, from the work of Tompkins *et al.* and Heal that there are two processes occurring in the decomposition; one requiring little or no thermal activation and the other a small amount. Both agree, that decomposition occurs at cation vacancies and involves N_3 radicals or an N_3 radical plus an azide ion. Heal established that no nitride ions are produced in the decomposition and the only products are metal plus nitrogen.

Boldyrev *et al.*[169] and Oblivantsev *et al.*[154] both find a serious discrepancy in the stoichiometry of the decomposition products and attribute this discrepancy to reaction of the free metal with air (or moisture, etc.). Hence some of the color bands observed (see reference 167) may indeed be due to this reaction of the metal atoms with air.

The observation by Oblivantsev *et al.* of a decrease in *G* value for the decomposition of the alkali azides with high LET radiation we believe is in error. The ratio of the decomposition yields of potassium, rubidium and cesium azides to that of sodium is about the same for the gamma ray and proton induced decompositions:

Ratio	Gamma radiolysis	Proton radiolysis
$G(-KN_3)/G(-NaN_3)$	1.64	1.52
$G(-RbN_3)G/(-NaN_3)$	2.52	2.42
$G(-CsN_3)/G(-NaN_3)$	4.15	3.9

This would indicate that there is no effect of LET on the decomposition.

5.2   Sulfates

The available information on the decomposition of the sulfates is very limited. Johnson[71] studied the decomposition of $FeSO_4\ 7\ H_2O$ and $FeSO_4(NH_4)_2SO_4\ 6\ H_2O$ using 2 MeV electrons at a dose rate of 0.63 $\times\ 10^{18}$ eV/g/sec. The products observed after dissolution of the irradiated salt in water were ferric ion, sulfite ion and hydrogen. The overall stoichiometry being

$$4\ Fe^{2+} + H_2O + 2\ H^+ + SO_4^{2-} \longrightarrow 4\ Fe^{3+} + H_2 + SO_3^{2-} + 2\ OH^-. \qquad (5\text{-}18)$$

The initial yields of products appeared to be linear with absorbed dose, however, at higher absorbed doses there was a definite decrease in G values for all the products observed. The product yields were the same whether or not ammonium ion was initially present. Initial G values were $Fe^{3+} = 1.75$, $SO_3^{2-} = 0.43, H_2 = 0.35$. It was originally believed that the H_2 arose from reaction of trapped electrons with the solvent, but subsequent studies[171] established that the hydrogen arose from decomposition of water of hydration. Dosimetry was not accurate in these studies hence the reported G values are only approximate.

Wigen and Cowan[172] examined the paramagnetic resonance absorption in $Li_2SO_4H_2O$ and $CaSO_4\ 2\ H_2O$ which had been irradiated with 1 MeV electrons. The crystals were irradiated and examined at 77°K. In lithium sulfate they observed two strong centers which were attributed to a hole localized on the hydrogen bonded sulfate oxygen and a trapped electron exhibing the free electron g value. The results for calcium sulfate were interpreted similarly.

Spitsyn, Gromov and Karaseva[173] examined the EPR spectra of sulfates containing radioactive ^{35}S and irradiated anhydrous alkaline earth sulfates. In radioactive strontium sulfate two signals were observed at $-197°C$. One a singlet with a g factor of 2.004 was attributed to the SO_4^- radical ion, the other center was attributed to a cation which has captured an electron (Sr^+). Samples of the sulfates which had been irradiated with cobalt-60 gamma rays gave spectra similar to those of the radioactive samples except that the signals were poorly resolved. In irradiated calcium sulfate, four or five overlapping signals were observed, some of the signals persisting up to 150°C. This stable signal was attributed to a cation which has captured an electron. The only other signal which was identified was that due to the SO_4^- ion.

Huang and Johnson[174] studied the decomposition of a number of sulfates in an attempt to obtain some information on the relative stability of the sulfate ion in various lattices. The salts studied were $MnSO_4 H_2O$, $NiSO_4 6 H_2O$, $CoSO_4 7 H_2O$, $ZnSO_4 6 H_2O$, $MgSO_4 7 H_2O$, $CrSO_4 15 H_2O$, $CuSO_4 5 H_2O$, and sodium, potassium, cesium and rubidium sulfates. Irradiation of the anhydrous alkali metal sulfates up to an absorbed dose of 4×10^{22} eV/g produced no detectable decomposition of the sulfate ion. For Na_2SO_4 no decomposition was observed at temperatures as high as 300°C. With the exception of $MnSO_4 H_2O$ and $NiSO_4 6 H_2O$ no decomposition was observed in any of the salts except that of the water of hydration. The G values for the decomposition of the water of hydration was appreciably less than that observed in ice, i.e., G values were all less than 0.01 for the hydrates whereas for ice it is known to be greater than 0.1. In $MnSO_4 H_2O$ the products of the decomposition were Mn^{3+} and H_2, however, the products actually detected were MnO_2 and H_2. The MnO_2 presumably arising from the following reactions:

$$\text{solid } 2 Mn^{2+} + 2 H_2O \rightsquigarrow 2 Mn^{3+} + H_2 + 2 OH^- \qquad (5\text{-}19)$$

the Mn^{3+} ion upon dissolution in water disproportionates

$$2 Mn^{3+} (aq) \longrightarrow Mn^{2+} aq + Mn^{4+} \qquad (5\text{-}20)$$

the Mn^{4+} ion subsequently hydrolyzing to give MnO_2

$$Mn^{4+} + 2 H_2O \longrightarrow MnO_2 + 4 H^+. \qquad (5\text{-}21)$$

The increase in acidity actually observed on dissolution of the irradiated salt in H_2O was consistent with equation (5-21). No decomposition of the sulfate ion in the manganous salt occurred which was to be expected since the manganous ions are tetrahedrally coordinated with sulfate oxygen atoms and water molecules. G values were dose dependent with initial G for H_2 being about 0.07.

The products of the decomposition of $NiSO_4$ $6 H_2O$ as determined by dissolution of the irradiated salt in water were comparable to those observed in the ferrous salt; i.e. Ni^{3+}, SO_3^{2-} and H_2 were observed in the ratio $4:1:1$. The product yields were dose dependent and reached a steady state at an absorbed dose of about 3×10^{21} eV/g. (See Figure 5-4). Decomposition at low temperatures did not produce any significant difference in the relative product yields.

A mechanism for the decomposition of the sulfate hydrates was given as follows:

INITIAL ACT

$$H_2O \xrightarrow{\text{\scriptsize\Large\leadsto}} H_2O^+ + e^- \tag{5-22}$$

$$e^- + H_2O \longrightarrow H_2O^- \tag{5-23}$$

$$H_2O^+ + M^{2+} \longrightarrow M^{3+} + H_2O \tag{5-24}$$

$$H_2O^+ + SO_4^{2-} \longrightarrow HSO_4^- + OH^- \tag{5-25}$$

$$HSO_4 + H_2O^- \longrightarrow SO_3^{2-} + OH^- + H_2O \tag{5-26}$$

or $$2 H_2O^- + SO_4^{2-} \longrightarrow [2 H_2O + SO_4]^{4-} \tag{5-27}$$

$$[2 H_2O + SO_4]^{4-} \longrightarrow SO_3^{2-} + H_2O + 2 OH^- \tag{5-28}$$

$$[2 H_2O + SO_4]^{4-} \longrightarrow SO_4^{2-} + H_2 + 2 OH^- \tag{5-29}$$

$$H_2O^+ + H_2O^- \longrightarrow 2 H_2O \tag{5-30}$$

$$2 (H_2O^+) \longrightarrow \tfrac{1}{2} O_2 + H_2O + 2 H^+ \tag{5-31}$$

$$(H_2O^-) \longrightarrow \tfrac{1}{2} H_2 + OH^-. \tag{5-32}$$

5.3 Carbonates

There have been no studies reported on the radiation induced decomposition of the carbonates except those related to paramagnetic species formed during irradiation. The species which have been identified in calcite crystals are: CO_3^{3-}, CO_3^- (both axially symmetric and orthorhombic)[175,176,176a] CO_2^-[177] and the radical $CaCO_3$.[178] Calcite crystals irradiated at liquid nitrogen temperatures emit an intense blue luminescence[178] indicating that electron hole recombination is occurring. This hypothesis was supported by the

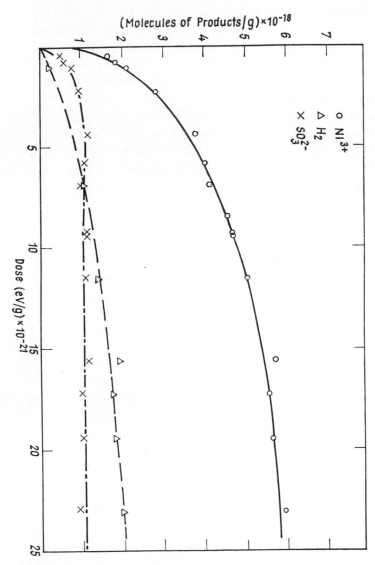

FIGURE 5-4   Yields of NiSO$_4$ · 6 H$_2$O decomposition as a function of absorbed dose

fact that the ESR signals for CO_3^{3-} and CO_3^- decreased a similar fashion to the luminescence. It was also observed that the ESR signals of these two centers were reduced in intensity 100 fold by 15 minutes exposure to xenon arc light filtered by Pyrex ($\lambda \sim$ 3500–7500 Å). Analysis of line widths indicated that the electrons (CO_3^{3-}) and holes (CO_3^-) were independantly trapped at well separated anion sites. An estimate of the G values for these paramagnetic centers was 4.0.[178]

It would appear that since CO_2^- is observed in irradiated calcite that some decomposition of CO_3^{2-} has occurred, however, no measurements of the decomposition have been reported. In irradiated potassium bicarbonate a signal attributed to HCO_2 has been reported[179] which would indicate that bicarbonates are also susceptible to radiation damage.

5.4   Permanganates

Although the available data on the radiation induced decomposition of the permanganates is very meagre, the data on the effect of pre-irradiation on the kinetics of the thermal decomposition of these salts support the fact that decomposition does occur.

Glemser and Buthenuth[180,181] observed the decomposition of potassium permanganate when exposed to 50 kV electrons in the beam of an electron microscope. They observed that MnO_2, which separates out in an independent phase, and MnO_4^{2-}, which forms a solid solution with permanganate ion, and oxygen are the major products. Evidence for the presence of MnO_4^{3+} ion was also obtained but this product appeared only after prolonged exposure to the electron beam. The overall decomposition may be represented as

$$5 \, KMnO_4 \xrightarrow{} K_2MnO_4 + K_3MnO_4 + 3 \, MnO_2 + 3 \, O_2.$$

Boldyrev *et al.*[182] have explained the accelerating effect of preirradiation on the thermal decomposition rate of permanganates as being due to the superposition of two factors working in opposing directions: (1) MnO_2 which catalyzes the thermal decomposition and (2) MnO_4^{2-} which acts as an inhibitor. There is no discussion, however, of how MnO_2 accelerates the decomposition or how the manganate ion inhibits it.

In general, it does appear that the permanganates undergo decomposition, however, the kinetics must certainly be very complex.

References

159 M. C. R. Symons, in *Radiation Chemistry* Vol. II Advances in Chemistry Series no. **82** Amer. Chem. Soc. (1968) Washington, D. C.

160 N. F. Mott, *Proc. Roy. Soc. A*, **172**, 325 (1939)

161 P. V. Gunther, L. Lepin and K. Andreew, *Z. Electrochem.* **36**, 218 (1930)

162 J. G. N. Thomas and F. C. Tompkins, *Proc. Roy Soc. A*, **209**, 550 (1951)

163 P. W. M. Jacobs and F. C. Tompkins, *Proc. Roy. Soc. A*, **215**, 254 (1952)

164 J. M. Groocock and F. C. Tompkins, *Proc. Roy. Soc. A*, **223**, 267 (1954)

165 F. P. Bowden and K. Singh, *Proc. Royal Soc. A* **226**, 22 (1954)

166 H. Muraour and A. Ertaud, *Compt. Rend.* **237**, 700 (1953)

167 H. Rosenwasser, W. Dreyfus and P. Levy, *J. Chem. Phys.* **24**, 184 (1956)

168 H. G. Heal, *Trans. Far. Soc.* **53**, 210 (1957)

169 V. V. Boldyrev, A. N. Oblivantsev and V. M. Lykhin, *Proc. Acad. Science S.S.S.R., Phys. Chem.* **159**, 1099 (1964)

170 Y. A. Zakharov, S. M. Ryabykh, and A. P. Lysykh, *Kinetics and Catalysis* **9**, 562 (1968)

171 E. R. Johnson, *J.A.C.S.* **78**, 5196 (1956)

172 P. E. Wigen and J. A. Cowen, *J. Phys. Chem. Solids* **17**, 26 (1960)

173 I. Spitsyn, V. V. Gromov and L. G. Karaseva, *Proceeding of the Academy of Sciences S.S.S.R.*, **159**, 1017 (1964)

174 S. Huang and E. R. Johnson, ASTM Symposium "Effects of High Energy Radiation on Inorganic Substances" Seattle, Wash. Oct. 31–Nov. 5, 1965 ASTM Special Tech. Publ. No. 400

175 R. A. Serway and S. A. Marshall, *J. Chem. Phys.* **46**, 1949 (1967) *ibid* **47**, 868 (1967)

176 S. A. Marshall, J. A. McMillan and R. A. Serway, *ibid* **48**, 5131 (1968)

176a R. S. Eachees and M. C. R. Symons, *J. Chem. Soc. A*, **790** (1968)

177 S. A. Marshall, A. R. Reinberg, R. A. Serway and J. A. Hodges, *Molec. Phys.* **8**, 225 (1964)

178 J. Cunningham, *J. Phys. Chem.* **71**, 1967 (1967)

179 G. W. Chantry, A. Horsfield, J. R. Morton and D. H. Whiffen, *Molec. Phys.* **5**, 589 (1962)

180 O. Glemser and G. Buthenuth, *Naturwiss.* **40**, 508 (1953)

181 *Ibid Optic* **10**, 42 (1953)

182 V. V. Boldyrev, A. N. Oblvantsev, A. M. Raitsimling and E. M. Uskov, *Proc. Acad. Science, USRR*, **166**, 891 (1966)

Index